Places, Affordances, At

This book offers a diverse understanding and practical approach towards the growing area of atmosphere research, in the context of philosophy, geography and architecture.

It begins by tracing back to the model of experience called the "pathic". Drawing on the phenomenology of theorists Hermann Schmitz and Gernot Böhme, introductory chapters offer a grounding for the beginnings of pathic research. The chapters go on to apply pathic framework to a range of practical cases from theatre studies to education. Atmospheres are often defined as affects one feels in a "lived space" and researchers are becoming more interested in the emotions we feel in natural and artificial environments across day to day life. By providing a critical re-evaluation of phenomenology and aesthetics, the book brings a series of unexplored and controversial subjects to light, opening up a new context for thinking about our everyday life and experiences inscribed within aesthetics, politics, literature, spatial practices and pedagogy and effectively merging abstract philosophy and concrete practice.

This book is particularly poignant in the emerging field of Atmosphere and New Aesthetics research. Practitioners, academics and researchers working within Cultural Geography, Aesthetics, Art and Philosophy will find this book extremely valuable.

Tonino Griffero is full professor of Aesthetics at the University of Rome "Tor Vergata", Italy. His most recent books (in English) are: *Atmospheres. Aesthetics of Emotional Spaces* (2014) and *Quasi-Things: The Paradigm of Atmospheres* (2017).

Ambiances, Atmospheres and Sensory Experiences of Spaces

Series Editors: Rainer Kazig, *CNRS Research Laboratory Ambiances – Architectures – Urbanités, Grenoble, France*
Damien Masson, *Université de Cergy-Pontoise, France*
Paul Simpson, *Plymouth University, UK*

Research on ambiances and atmospheres has grown significantly in recent years in a range of disciplines, including Francophone architecture and urban studies, German research related to philosophy and aesthetics, and a growing range of Anglophone research on affective atmospheres within human geography and sociology.

This series offers a forum for research that engages with questions around ambiances and atmospheres in exploring their significances in understanding social life. Each book in the series advances some combination of theoretical understandings, practical knowledges and methodological approaches. More specifically, a range of key questions which contributions to the series seek to address includes:

- In what ways do ambiances and atmospheres play a part in the unfolding of social life in a variety of settings?
- What kinds of ethical, aesthetic, and political possibilities might be opened up and cultivated through a focus on atmospheres/ambiances?
- How do actors such as planners, architects, managers, commercial interests and public authorities actively engage with ambiances and atmospheres or seek to shape them? How might these ambiances and atmospheres be reshaped towards critical ends?
- What original forms of representations can be found today to (re)present the sensory, the atmospheric, the experiential? What sort of writing, modes of expression, or vocabulary is required? What research methodologies and practices might we employ in engaging with ambiances and atmospheres?

Atmospheres and the Experiential World
Theory and Methods
Shanti Sumartojo and Sarah Pink

For more information about this series, please visit: www.routledge.com/Ambiances-Atmospheres-and-Sensory-Experiences-of-Spaces/book-series/AMB

Places, Affordances, Atmospheres

A Pathic Aesthetics

Tonino Griffero

Routledge
Taylor & Francis Group
LONDON AND NEW YORK

First published 2020
by Routledge
2 Park Square, Milton Park, Abingdon, Oxon OX14 4RN

and by Routledge
605 Third Avenue, New York, NY 10017

First issued in paperback 2020

Routledge is an imprint of the Taylor & Francis Group, an informa business

British Library Cataloguing-in-Publication Data
A catalogue record for this book is available from the British Library

Library of Congress Cataloging-in-Publication Data
A catalog record has been requested for this book

ISBN 13: 978-0-367-72761-1 (pbk)
ISBN 13: 978-1-138-38937-3 (hbk)

Typeset in Times New Roman
by Taylor & Francis Books

Contents

Preface

Perhaps pathicity does not perfectly coincide with what is "other than reason", as it is often emphatically defined (cf. e. g. Böhme and Böhme, 1985). However, precisely because it respects this "other" as it appears to us, free from the Procrustean bed of transcendental bonds, the notion of pathicity certainly does not perfectly coincide with the Enlightened reason either, which is induced to abruptly degrade, *à la* Hegel, the irrational to the unreal. Philosophically rehabilitating pathicity thus means giving value – in a non-academic sense – to letting oneself go, to being "*subject to*" rather than "*subject of*". This amounts to defending the unfashionableness, and for this reason the critical relevance, of the gift of knowing how to be a means of what happens to us rather than of doing – something that today is very rare indeed. Promoting abandonment may appear as a performative contradiction (one should get busy in order to be passive), but it really is not the case. The hyper-rationalist dogma of full subjective self-determination, corroborated by the idea that action comes first and that the meaning of life only lies in planning one's goals, cannot share and perhaps not even understand such a skill.

According to Ludwig Klages,[1] pathicity is the ability to receive effects rather than produce them, to be possessed by what happens and to act only in light of this rapture. More precisely, it consists in living and contemplating the images active on us, which are real *Urphänomene*, at a distance and outside of consciousness (see, e.g., Klages, 1929–32, p. 520). For Klages, only what is independent of consciousness – that is, an almost vegetative pathicity of the soul and of rhythm – is truly real. Thanks to these images, human beings today are philosophically immunized, at least partially, against the universal myth of mediation and politically immunized (in a broad sense) against the proximal standardization of seeing[2] and against the artificial teleologization imposed by modernity.

A less metaphysical and *kulturkritisch* view of pathicity can be found in Erwin Straus.[3] After suggesting that we pay attention not so much to sensations but rather to feeling as such, by "pathic" he simply indicates the moment of every perception that happens "at the most original stage of lived experience", but is unfortunately usually underestimated to the benefit of the gnostic moment. Therefore, the immediate, sensitive and preconceptual communication we have with appearances and their ever-changing presentation is pathic. Pathicity is particularly influential in the so-called lower senses, like touch, and in everyday habits, but also in the body,[4] when the eyes become less important than the torso (as in dance). However, for Straus, pathicity is also present in so-called higher senses, and even in seeing. In all perceptive experiences, in fact, and not only in exceptional ones or in those exceeding a certain perceptive threshold, one is always emotionally struck in a certain way: that is, one is involved not so much by what is given but rather by *how* it is given.

What I will discuss in the pages to follow is precisely this "how" of the lived experience of presence (see *infra* 8) and of being grasped, which we tend to react to in a rather regular way, despite the claims of gnostic-rationalistic ocularcentrism.[5] Of course, here I am talking about pathicity as opposed to both patheticity, with its sentimentalist patina, and pathology, with its physicalistic normativity. Nor am I trying to promote the notion of "pathos for its own sake" (*dolorismo* as supposed salvific suffering). Rather, I will focus my theoretical attention on the affections experienced as rapture and felt-bodily involvement, which one cannot critically react to nor escape from.

The event of being-grasped is essentially different from the act, both because it cannot coexist with the spontaneous initiative of the (same) subject, and because it is often in contrast with what the latter knows. As an extreme case of receptivity and openness to the other from us, being-grasped is a state, like pain, where it makes no sense to differentiate between feeling and felt. *Pathos* does not in principle imply a given subject, because it precedes subjective passivity and activity[6] as well as dualism. The latter, instead, is required both by a third-person cognitive approach, and by the ethical need for an (allegedly) autonomous self able to control their pathic states just as they do with their other psychic activities. Replacing the Cartesian model of reflection with the idea that thoughts can "inhabit us" without being objects of the reflection of the "I think", it can be said that such affective grasping is *not something that one does, but something that one is*. Or, better, it is not that *I* am affected by something, but rather that something affects *me*, where this "me" – in short, a dative or an accusative[7] implying that what happens *to me* affects *me* personally – acts as the pathic-aesthesiological *fundamentum inconcussum*. As absolute familiarity with itself, it replaces an "I" that, in Cartesianism, is always something potentially reflective and objectifying, something juridically and morally imputable, as well as something very claustrophobic.

The unfreedom that comes with giving oneself to something that cannot be the object of representation, however, should not be confused with inertia. It is a passivity that is so strongly operative as to often border with passion – in any case, it is a skill that must be painstakingly rediscovered in an age clearly obsessed with constructivist blinders. It is a pathicity that, while undoubtedly referring to something that is outside of us, cannot be domesticated through the rules imposed by the general alibi of intentionality (especially if understood as intentionality of acts rather than as pre-reflective operating intentionality), nor does it coincide with the ex-centricity that Plessner claimed as a mark of anthropic primacy. In other words, it is something that happens (to us) regardless of what we want, lasting as long as the emotional authority of what happens to us; it is something that we do not know,[8] but that we *feel*, and of which our felt body (*Leib*) is the faithful sounding board.

Note well: the pathicity I refer to here does not necessarily imply the prophetic cosmicity that animated, with sometimes even caricatural outcomes, the bohème of Stefan George's circle in Munich, notoriously fascinated precisely by raptures of various nature (*Ergriffenheit*). More simply, it is the acknowledgement that the aesthesiological receptivity that mainly characterizes our involuntary vital experiences and indirectly also the cognitive sphere – especially when it is involving to the point of "cornering" the perceiver (who is therefore "pre-occupied" with it) – still requires an adequate philosophical thematization.[9] This desirable reflection actually takes the connotation of a real *counter-discourse* [10] set against both the modern dogmatic conception of the subject (albeit in the direction of a pre-personal subjectivity as opposed to the fictitious one of the Es, as we shall see) and against analytic reductionism, responsible for creating irreparable divisions in the experience of life. Its subsequent and possible synthesis, as noted by Goethe in his conflict with the supposed improvement of any re-construction of the given, inevitably entails the loss of the original phenomenon.

Of course, there are some difficulties. In the first place, identifying itself with the feeling subject and escaping the representative objectivization that reduces it to a variant of the Cartesian I, how can pathicity be philosophically, and therefore reflectively, thematized? Also, a pathically involved person cannot wonder about the meaning or value of their involvement: at most they can react to it on the basis of their equally pathic felt-bodily dispositions, because by reflecting on it they would *ipso facto* admit not to be involved in it. In other words, *logos* and *pathos* cannot coexist in the same subject at the same time, but they can (indeed, must) alternate, thus giving life to a less neurotic form of life precisely because it is open to both options, more precisely to a form of existence that is able to reflect on what happened only because it was previously fully involved by it. *A fortiori*, as Merleau-Ponty emphasizes, a philosopher must reflect on the unreflected, albeit with the (neo)phenomenological caveat of sticking as much as possible to daily experience and, in the context of an aesthetic experience (in the broad sense), to a perception understood as a feeling alien to the cognitive, ortho-aesthetic and ultimately tantalizing temptation often associated with the perceptological task.[11]

A pathic aesthetics,[12] far from presupposing an interpretive and constructivist approach to the world, supposes that there is a meaning already sedimented outside of us. Therefore, its objective is to theoretically outline the contours of a sphere animated by feelings that are effused in the pre-dimensional space (atmospheres) and so quasi-thingly aggressive as to take possession of us (maybe even only hypodermically). This sphere appears necessarily and structurally vague (only as certain as an *experientia vaga* like the aesthesiological one), but it is still essential to give our lives the "colour" that, alone, makes it meaningful or at least tolerable: only feeling certifies our existence and the reality of the world – that is, the active one (*Wirklichkeit*), and not only the thingly one (*Realität*).[13] The aesthetic "justification" of our lives, therefore, does not lie in works of art, in which – albeit thanks to an affective-somatic medium – the po(i)etic product prevails over the pathic, but in the much more salient (and possibly contagious) everyday feelings.

But are we really able to take what happens to us in a casual and ephemeral way, without resorting to regressive principles or getting lost in late-Romantic nostalgias? Are we able to resist the temptation of necessarily transforming the "given" into something "done", in the illusion that everything can be "produced"? Can we accept, in the name of a phenomenology that preserves the affective-atmospheric given, the necessary role that pathicity[14] and contingency play in our lives? This thought of the senses (where this genitive is both subjective and objective) is able to respond to the "constraints arising from the lived experience of reality" (Klages, 1929–32, p. 1419), and to those feelings that – unlike thoughts (free, available and ideally infinite) produced by intellect and fantasy – are ultimately always the same, already mature at their first appearance as, almost like daemonic forces, they exist without being available to us or to our control.[15] Can we really consider such thought perfectly rational? As pointed out by Carl Haensel, thoughts are born in us, they are countless like sand grains in the sea, their content is inexhaustible and variable. Feelings take us over in the form of a power that comes from the outside, they grasp us, they attack us, and we are possessed by them. In principle we can think of anything, even the absurd, things that never existed and never were realized, things with no meaning. On the contrary, the feelings we are capable of can be counted on the tip of one's fingers (Haensel, 1946, p. 52).

Notes

1 Klages (1974, p. 643; 1964, pp. 167, 422) thinks he reintroduced pathicity to modern thought (cf. Preusser, 2015, pp.143 ff.).

2 It could be said, half-jokingly, that "television (as a device and as technology) destroys television–viewing (as pathic contemplation)" (Preusser, 2015, p. 157).

3 Cf. Straus (1963, pp. 367–379 and passim, 1980, p. 11 ff.) but also Warburg's (perhaps too) famous notion of *Pathosformeln*.

4 Unfortunately only in terms of physical (*Körper*) and not felt (*Leib*) bodiliness (Straus, 1963).

5 In my opinion, Straus's view (1963) that observation objectifies and universalizes while gesture avoids this depersonalization goes well with Klages' idea (which Straus refers to). By articulating the body into mainly pathic parts and mainly gnostic-practical parts, Straus then paves the way to the neo-phenomenological division of the felt body into quasi-thingly "felt-bodily isles" to which I will come back many times (but cf. Griffero, 2017a, pp. 55–67).

6 "One who suffers from something painful does not wonder, as a sufferer, what is causing one's suffering and whether this is objectively justified", as in that situation one lacks "any distance and any subjective distanciation" (Bernet, 2014, p. 19).

7 For a philosophy of "to me" cf. Böhme (2012, pp. 11–22), while, for a perception that refers not so much to an "I" but to a "me", see Wiesing (2014).

8 Provided, of course, that a situation that is pathically experienced in the first person cannot be also called knowledge.

9 This theme is implicit both in Schmitz's notion of affective involvement (*affektive Betroffenheit*) and in Böhme's claim of an oblique thought, by virtue of which the subject would recognize their own dependence and dissolve the narcissistic image they otherwise have of themselves (Böhme, 2012, pp. 188–195).

10 Opposed to taboos aimed at excluding from reason anything that is traditionally an obstacle to it, for example the unconscious, the felt body and nature (Cf. Böhme, 2012, p. 22).

11 As noted by Waldenfels (2000, pp. 97–98).

12 For an overview, cf. Griffero (2017a, pp. vii–x).

13 "We are not alive because we think, but because we feel". And, moreover, "we do not feel in order to live, but we live for the feeling" (Haensel, 1946, pp. 11, 298).

14 Whether it is the physical and felt body that we have and are, the language we speak, the era in which we live, the tradition to which we belong, our "destiny" or, if you prefer, our biography (Böhme, 2014, p. 147).

15 According to this theory of feelings as daemonic-mythological entities (cf. Haensel, 1946, pp. 38–39, 41, 48–49, 302, 309–310), they are objective and cosmic realities.

Author's note

The chapters of this book are based on texts that have been presented at various conferences and workshops, and that appear here in a revised (expanded or reduced) form. I thank the publishers for allowing me to translate and/or publish them here.

The Preface and some chapters (1, 2, 3, 4, 10, 12, 13, 14) are the translation of some paragraphs (1.1, 1.2, 1.3, 1.4, 1.5, 4.1, 4.2, 4.3, 4.8, 4.9, 2.9, 3.4) of my book *Il pensiero dei sensi. Atmosfere ed estetica patica*. Milano: Guerini & Associati, 2016. They have undergone several revisions and have been expanded (or reduced) specifically for this book, on the basis of valuable feedback I received.

Chapter 5 is based on Falsi sentimenti (atmosferici)? Autentico e inautentico nella sfera emozionale. In: Catalano, G., Ciccarini, M., Marcialis, N. (eds), *La verità del falso*. Studi in onore di C. G. De Michelis. Roma: Viella, 2015, pp. 143–150.

Chapter 6 comes from Buone regressioni. Meità e impersonalità alla luce della Nuova Fenomenologia. *Philosophy Kitchen* 3, 5 (2016), pp. 126–138.

Chapter 7 was developed based on Condannati al senso (e all'espressione). Otto tesi sulle atmosfere come insiemi espressivi. In: Benenti, M. and Ravasio, M. (eds), *Espressività. Un dibattito contemporaneo*. Milano: Mimesis, 2017, pp. 169–195.

Chapter 8 is a revised version of Come rain or come shine… The (neo) phenomenological will-to-presentness. *Studi di estetica* XLVI, 4, 11 (2018), pp. 57–73.

Chapter 9 is based on a text published as Architectural Affordances: The Atmospheric Authority of Spaces. In: Tidwell, P. (ed.), *Architecture and Atmosphere*, Espoo: Tapio Wirkkala-Rut Bryk Foundation, 2014, pp. 15–47.

Chapter 11 is the revised translation of Inclusività atmosferica. In Longo, A., Rabbiosi, C., Salvadeo, P. (eds), *Forme dell'inclusività. Pratiche spazi progetti*, Santarcangelo di Romagna (RN): Maggioli Editore, 2017, pp. 31–51.

Chapters 15 and 16 have been written specifically for this book.

Part One

Pathic aesthetics and New Phenomenology

1 Art and atmospheres

The approach of a pathic aesthetics

If aesthetics were really "only" a philosophy of art, it would not be far-fetched to consider it no longer necessary today.[1] Indeed, it would be a form of reflection that – after reaching its apogee in the 19th century on the background of an upper-bourgeois culture inclined to see the art world, considered autonomous by definition, as an authentic surrogate of faith – may have largely lost its *raison d'être.* Not so much because, as posited (if taken literally) by the famous Hegelian "obituary", the Christian unrepresentability of the purely spiritual makes art something irreversibly past within the absolute spirit. Rather, because from the standpoint of an interpretation of the felt-bodily presuppositions of the whole historical-cultural evolution, the stylizing creative force might turn out to be weak in the aftermath of the Romantic age, due precisely to a collective disposition of the bodily experience that has become more and more superficial.

But this diagnosis *could* perhaps apply – it is mandatory to use the conditional – to an aesthetics whose only theme was art, while it certainly does not seem to affect aesthetics in a wider sense. In fact, it does not prejudice at all an aesthetics that re-evaluates its 18th-century genesis (Baumgarten) and is conceived rather as a philosophy of sensible knowledge (or sensibility tout court), or as a *thought of the senses* (in both the subjective and the objective sense of the genitive). Thus conceived, in fact, aesthetics survives without particular anguish its somewhat megalomaniac phase where it claimed to see art and its unspecified masterpieces as many different things: the revelation of an absolute and an identity that are otherwise only conjectural and interior (Schelling); the sensible manifestation of the Idea (Hegel); the place where the truth of a Being strategically concealed in favor of simple presences that are widely available and thus ontologically disqualified "sets itself to work" (Heidegger); and perhaps the anticipation of utopian social conditions aimed at compensating humans for the wrongs inflicted on them by history (Adorno, Bloch).

Thus released from this view of art as a continuation "with other means" of religion and/or politics, aesthetics seems finally able to (once again) strongly claim its role as the – not gnostic but pathic (see the Preface) – philosophy of felt-bodily feeling,[2] rehabilitating typical suggestions of 18th-century "effect"-

based aesthetics which lost momentum in the 19th century due to idealist *top-down* aesthetics (cf. Fischer-Lichte, 2005, p. 41). As the pathic philosophy of felt-bodily feeling, aesthetics can even aspire to the role of "first philosophy", using art – at most – as an emblematic example. On the one hand, this discipline rejects the otherwise irreversible process of "cerebration" (Gottfried Benn) characteristic of Western rationalism, being concretely committed to rebalancing its one-sidedness; on the other hand, it aims to combat the immunization from the problems of the present that was (more or less voluntarily) enacted by the standard aesthetic theory (cf. Böhme, 1995, p. 11). Therefore, the point is to shape the aesthetic reflection no longer on social and communicative models, largely indebted to the implicit Kantian bias of a disinterested distality, but on the pathic – therefore sensible and felt-bodily – dimension which innervates life as a whole, including the most futuristic technoscientific opportunities.

But this contagious and destabilizing involvement, perhaps widely favoured by communicative globalization and basically only emphasized by contemporary "immersive" art,[3] must now be framed within the wider paradigm of so-called New Phenomenology.[4] In full harmony with (if not anticipating) the current rebirth of the multidisciplinary interest in emotions – but without being subject to neuroscientific hypotheses, the superficial mimetic forms of an extrinsic *pathos* or the "immersive" exaggerations of contemporary media studies – this heterodox phenomenological current sees philosophy as a reflection on how "one feels" or "one finds oneself" in a certain space (obviously not geometric but lived and pre-dimensional) and in the presence of involuntary vital experiences. But above all, it rejects the total psychic introjection of affective states, considering them, *qua* atmospheres (as we will see), external to the subject and often for this reason refractory to any subjective projection. The relatively predictable aesthetic consequence of this premise will then be first of all to investigate not so much art and the traditional categories of aesthetics (beautiful, sublime, artwork, grace, genius, originality, mimesis, etc.), but rather the implicit pathic structure in any regional ontology (including art, of course). The second goal will be to limit as much as possible the intellectualistic level of critical judgment (especially the one focused, as usual, on art), so as to start reflecting on the general aesthesiologic potentialities of (especially "naive") perception.

If it really wanted to focus on art, a pathic aesthetics could simply recognize it as one of the most significant ways in which the human being, making the affective more actively real, is invaded by it "without the risk of concrete reality" (Haensel, 1946, pp. 365, 354); art would be the place where we make our emotional turmoil acceptable and therefore manageable. At most, art is therefore the historically specific modality with which the European intellectual, in an era interested in the production of instrumental objects – but also of experiences exempt from pragmatic urgencies, and therefore automatically considered symbols of freedom and even of sacredness – enjoyed a sort of "trial feeling" (Böhme, 1989, p. 153). Art can thus be considered a form of domestication and "cultivation", within a controlled space, of atmospheres diffused in our "surroundings", of "daemonic" unmanageable feelings to which the percipient can react, according to their own felt-bodily disposition, either by giving themselves fully to them or by escaping them (even physically).[5] Such spatialized feelings are situations whose internally diffuse meaningfulness is formed by a chaotic – and therefore atmospheric[6] – manifoldness that cannot be reduced to discrete (let alone linguistic) elements. These situations grasp us from the outside; therefore, in the experience of art we are able to keep our composure, re-elaborating and objectivating in a relatively stable form the felt-bodily and centripetal aggressions of atmospheric sentiments.

Be they linguistic or not,[7] works of art only bring to light atmospherically intrusive situations through veritable *Pathosformeln* (to borrow Warburg's term). The interior of a church imposes on those who enter it – whether or not they know its meaning – a certain emotional tone and a consequent behavior (of devotion or at least reverence); in the same way, a fugue by Bach, through motor suggestions and synaesthetic characters, radiates penetrating albeit indeterminate and almost numinous atmospheres (in Rudolf Otto's sense). In both cases, in fact, atmospheric feelings hovering in sufficiently circumscribed spaces grasp the felt body of the percipient, whose feeling, by virtue of a permanent and often infallible felt-bodily communication with the surrounding forms, is thus perfectly matched (attuned) to those feelings. In other words, the sets of expressive – physiologically tertiary but phenomenologically primary – qualities (see Chapter 7) which we legitimately call atmospheres find perfect *resonance* [8] in the perceiver. Indeed, atmospheres are anything but the outcome of associationist processes or subjective projections on the external world, which only appears to be a-qualitative, in reality, because it has been previously reductively emptied of its own immanent qualities.

To oversimplify, artistic forms are here a "bait" through which to capture atmospheres that are otherwise refractory if not openly destructive, and in any case endowed with authority even in this "tame" version. The experience we make of art, therefore, is frankly irreducible to the over-intellectualistic and disciplined psycho-gymnastics of the Kantian free play of the faculties. Let's take a few examples. In everyday life, the atmosphere of someone's severe sadness normally arouses respect; a euphoric atmosphere infects and carries away even the most reserved people; the centripetal atmosphere of shame imposes tightness, making the person responsible for it, or even an innocent spectator, occupy, so to speak, less space than what rightfully belongs to their body;[9] the atmosphere of injustice demands direct (or legal) retaliation from its victims. Well, in the case of the work of art, this authority or authoritativeness instead assumes obviously mitigated and contemplative forms. Perhaps this is the correct interpretation of the all-too-quoted "you must change your life" expressed, according to Rilke, by the archaic Torso of Apollo: perhaps it is indeed an affective and felt-bodily injunction addressed to the observer by a successful (albeit mutilated) art form.

The affordance suggested by a work of art therefore never consists in an entirely irrational turmoil. The rapture that it arouses suggests the regression to primitive presence – to which, according to Schmitz, it is absolutely healthy to regress from time to time, to ensure that one's identity is not purely formal and self-attributed but on the contrary absolutely subjective (*qua* affective and felt-bodily). However, the rapture is also an important phase of personal emancipation,[10] realized both by singularizing the initially undetermined situational semantic halo into a "this" (subsumption under the genus and species of art), and through the somewhat devotional "distance" emanating from the aesthetic orientation. Exactly like other privileged affective experiences (above all dwelling: see Chapter 10), art makes the observer be carried away by the influence of atmospheric impressions hovering in the environment but condensed in some pregnant objects, and yet it also makes sure that the observer does not sink into the artwork's atmosphere until they are completely subjugated by it. The balance thus achieved, resulting from a fortunate felt-bodily alternation between contraction and expansion, immersion and emersion, is nothing but the well-known state of quiet and recollection that characterizes every final cathartic condition: it is the condition of distanced reverence that circumscribes and, in a certain sense, crystallizes in the specific configuration – be it the work or the single aesthetic experience – the daemonic powers that follow each other in the lived space and in which, whether we like it or not, we are called to participate.

At this point, one can find numerous unexpected analogies between the pathic structure of aesthetics and that of law. Let's leave aside the most obvious analogies, such as the coherence, order and symmetry ("style") indispensable both in the work of art and in the legal system. Another unquestionable point of contact lies in the idiographic anti-deductivism inherent in the aesthetic judgment as well as in juridical assessments: while demanding an always creative hermeneutical application of the general to the particular, both law and art take the particular case (the crime and the beautiful) as the starting point of a fruitful reconsideration of the "prejudices" (in Gadamer's sense) that act as their initial parameters. Rather, what interests us here is the analogy between aesthetics and law (suggested by Hermann Schmitz) in terms of domestication of the pathic-daemonic, that is to say of those feelings that, while being unquestionably "the most important things in life, since only they bring force and delicacy, splendour and opacity in the world, and all that in general has true importance for man" (Schmitz, 1969, p. xiii), are often also destructive.

Like art and aesthetics, the law – obviously analyzed here regardless of any disciplinary technicality or any dubious imposition to reject the affective in favour of (anaffective) "facts" – is presented in this perspective as a distanced form of rapture and cultivation of external and aggressive atmospheric feelings. Now, there are of course some prejudices at play here. On the one hand, there is juridical positivism, which, being deprived in its formalism of the "*pathos* conferred to it [to the law] by the nimbus of legitimacy" (Schmitz,

2012, p. 42),[11] only justifies punishment with formal conscientiousness. On the other hand, there are the theories of eternal values (it makes no difference here whether they refer to God or to nature), whose binding validity remains fatally ideal and in any case always subject to doubt. However, if one looks beyond such prejudices it becomes finally possible to rediscover the affective roots of the very idea of justice, that is, of its indispensable and relatively autonomous nucleus with respect to the accessory and merely organizational components of social life.

In this way, one can acknowledge the powerful way in which the law involuntarily grasps those who are affectively touched by it, persuading them of being in the right not only in a conventional sense, but also because of a feeling of evidence. The validity of this feeling (which is anything but merely contemplative, cf. Schmitz, 2012, p. 28 ff.) is proven by the fact that they cannot avoid it at will and without guilt, or at least, they can hardly put forward arguments against it based on further critical reservations. What enjoys this feeling of evidence, thanks to the felt-bodily resonance it involves – which is far more intense than the ontological evidence leading to define a state of affairs as a "fact"[12] – is mainly the sense of injustice. The latter is indeed an irreducibly subjective "fact", without which everything would end up appearing indifferent, any arbitrariness would be passed off as legal and, at best, norms would remain ineffective.[13] So evident and therefore authoritative as to limit the critical freedom that otherwise would exempt one from obedience,[14] the feeling of injustice is present wherever something "feels" like it cannot be tolerated (think of the ontogenetically prototypical and pathogenic case of a child unfairly accused!). It involves the perceiver through anger (if they are suffering the injustice and feel they are in the right) and shame (if they are responsible for it and feel like they are in the wrong).

While admitting that there may be quite different thresholds of tolerability of injustice, the law as an abstract approximation to life[15] comes into play precisely to regulate this constellation of atmospheric feelings, and in particular to cultivate and thus make tolerable the quasi-thingly aggression of anger and shame. The latter are not only impersonal primal juridical feelings[16] somehow ontogenetically prior to good and evil, but they do not even depend on the availability of the legislator, so that their absence would make the punishment of some abuse as absurd as the punishment of an illness or a flooded river (Schmitz, 2012, pp. 60–61). Despite the different reactions they arouse – respectively compensation or revenge (in the more extroverted European law)[17] and annihilation of self-esteem or self-destruction (in certain more introverted archaic or oriental cultures, notoriously, up to suicide or harakiri) – anger and shame are perfectly complementary in their development.

In fact, anger towards the guilty is mitigated by the shame felt by the guilty, and, on the other hand, it is not rare to react with anger to shame (think of the duel, real or metaphorical). But these two primal sentiments become legal in the proper sense[18] only when there is a (strictly legal) presentiment of respect to support them, which, precisely because it presages the threat represented by anger[19] and shame, knows how to avoid it. It does so mainly without focusing on the *something* or *someone* which are respectively the sphere of condensation and the anchoring point of atmospheric anger and shame. Also, it does so by transcending these two cathartic feelings into a successful legal form: in the same way, the aesthetic presentiment (taste, tact) aims at the "creative" control of danger (atmospheric aggression) by virtue of a formal detour, or etiquette transforms brute desire into style and decorum, or else courtesy channels and refines potential anger in front of inconveniences caused by others.

Thanks to their peculiar "flair", aimed at avoiding respectively the ugly (or the unpleasant) and the unfair, aesthetics and the law develop largely analogous communities and institutions in their commitment to asymptotically get closer to the atmospheric authority to which they obey by means of (normative) practice. In these communities – with a further analogy – a juridical norm and an aesthetic judgment, initially binding only for a narrow circle, then try to impose themselves outside of it, usually circumscribing their claim of diffusion and consent only to the events and artworks that are presently less controversial. The juridical culture and the aesthetic culture that derive from it are however subject, albeit to a different extent, to a gradual revision of the respective cathartic forms, that is to say both of the lawsuit,[20] intended as purification of the initial and otherwise uncontrollable atmosphere of anger, and the execution or exhibition of works, intended as a way to test, in front of critics and public, proposals that often risk to provoke anger and/or shame for their sought originality (and, today, explicit provocation). In short, both cultures seem to guarantee the transition and authentic rite of passage from threatening and initially unmanageable atmospheres (pre-cathartic anger and shame) to tolerable and somewhat pacified atmospheres (juridical peace and critical consent).

In spite of these and other possible analogies between aesthetics and law, it seems difficult to imagine, except perhaps in the marginal *cupio dissolvi* of aestheticism, the existence of a Michael Kohlhass of beauty, that is of a feeling as seriously binding as that of injustice. The aesthetic feeling is not always so soft and tolerant[21] as to merely discuss why a mood is or is not shared, as can be already seen in the ideal stigmatization inherent in the appeal to universality in Kant's reflective judgment and then above all in the very concrete socio-cultural stigmatization perceived in taste as a segregating feeling of "distinction" (Bourdieu, 2010). However, the aesthetic feeling is undoubtedly less intransigent than the law, even within pathic aesthetics. Where beautiful and ugly express a condition of felt-bodily rapture characterized, respectively, by a relief oriented to privative expansion (wideness, even only as a promise) and by pressure oriented to privative contraction (tightness),[22] instead of punishing dissent, each person should simply try to persuade those who feel a different sentiment, inviting them above all to make the same experience (in the hope that within certain terms it will indeed be the same).

The beautiful and the just, however, as aesthetic and legal absolutes, while atmospherically tonalizing the space of those who get close or inside them,[23] almost always involve the perceiver only relatively. This is because, in the absence of a universality legitimized by logical rules, their authority rests on pathic sources mostly of a perspectival type, whose validity depends both on the felt-bodily resonance of the individual and on the latter's ability to critically transcend the involvement. For example, we are no longer ashamed of a grammatical error when it appears such to us only from a petty-bourgeois hyper-corrective perspective; similarly, we do not abandon ourselves to an undoubtedly persuasive melody when we notice its blatantly sentimental goal. Only rarely, when their evidence can no longer be relativized, do the beautiful and the just have absolute validity and authority for us: it is the case with "burning" shame, so centripetal and unquestionable as to compel assent, or with the "seductive" and literally uncritical pleasure aroused in us by an artistic or natural form whose contemplation we can no longer do without, in a certain sense. Normally, aesthetics and law rather generate a state of lucky mediacy between extreme and lethal cases represented on one side by emotional breakdown and on the other by the hyper-rationalization of existence, that is, between full emotional passivity and the obsessive and resentful control of feelings that involve us. Thus, they allow us to enjoy without risk the benefits of a rich and colourful pathic perspective, whose necessary thrill,[24] therefore, does not necessarily imply the regression or transformation of the percipient into a mindless and unwary blind passenger of freely floating atmospheres.

But what is an atmospherological aesthetics, and what does it look like? First of all, it should be noted that the aesthetics of atmospheres, which today rests on an already vast interdisciplinary literature, though not primarily oriented to art, seems paradoxically the only discipline able to provide an adequate understanding of contemporary art. In fact, the latter, in its adventurous quasi-thingly instances – ephemeral works, purely optical installations, interventions on nature that are only abstractly or photographically perceptible, intentionally extrasemantic synaesthetic constructions, perceptive dislocations, etc. – is notoriously more and more interested in generating intensely immersive atmospheres that involve the spectator in performative and not just contemplative practices. The same avant-garde metamorphosis of life into a total work of art (*Gesamtkunstwerk*) turns out to be possible and artistically manageable, in the final analysis, only if the immersive involvement (whose scope should never be exaggerated) is regularly followed by a semi-reflective distance. In any case, the point is not to insist on the relevance of an atmospherological aesthetics, if only because any work of art, *moins un monde qu'une atmosphère de monde* (Dufrenne), is ultimately only atmospheric because it selects and intensifies extra-artistic atmospheric impressions (cf. Hauskeller, 2002, p. 180). The latter are then conveyed on the basis of formal principles and of a social legitimation (museums, concert halls, publishing, etc.) to whose *atmosphere* – in another sense of the notion, namely the "institutional" one – one can sometimes refer in order to distinguish the work of art from the ordinary artifact. In short: I do not all seek to atmospherologically colonize the entire discourse on art, because the surplus, if present, is always twofold. In fact, while art in its rich socio-symbolic stratification is irreducible to an (only) atmospheric analysis (cf. Griffero, 2014a, pp. 82–87), atmospheric situations are irreducible to the *numerus clausus* of the very special, and far from eternal, things we call works of art.

But it is time to move beyond art, since the pathic aesthetics which I am trying to outline here is rather focused on sensible-affective perception in general. Because of this, it proves able to understand the (so-called) "no-longer-fine arts" (Marquard) in their asemanticity, being naturally interested in an emotional fruition that has little to do with the static-contemplative or reflective-judicative attitude. Also, and above all, it is able to understand today's lifeworld aestheticization, in all those activities (scenography, advertising, landscape planning, event organizing, museum set-up, cosmetics, urban planning, fashion, political and institutional communication, entertainment industry, emotional and multi-sensory marketing aimed at creating customer loyalty, interior design, etc.) which, once stigmatized by the elitist aesthetics of art, were legitimized by Gernot Böhme – also philosophically – with the name of *aesthetic work*. Therefore, my pathic aesthetics is mainly addressed to that sort of widespread design that influences us more than what we are willing to admit, and for which, with our every choice, we are also at least co-responsible.

Having to face problems that used to be at least partially unknown – from the ecological need for a renewed relationship with nature to the practical-theoretical need to be critically ready for the so-called experience society (*Erlebnisgesellschaft*) – the pathic and neo-phenomenological approach is a very ambitious project.[25] First of all, it is opposed to the radical scientification of experience, based on an (always and only) technologically mediated perception, and claims the inevitability of ordinary and affective (naive) experiences, of which one could say that they "just happen, and people do not consider how they happen, whether they could happen differently, or of what larger experiential structures they might be a part" (Seamon, 2000, p. 161). It is therefore only natural that this aesthetics should want to emancipate itself, as said, from its idealistic interpretation as a philosophy (or even metaphysics) of art, in order to return to (Baumgarten's) original project of aesthetics as general science of sensible knowledge (*aisthesis*).

Idealistic in focusing on the work and its supposed autonomy, metaphysical in considering art and beauty *sub specie aeternitatis*, bourgeois in its full adherence to the process of civilization (through abstraction) of the European elite, intellectualistic in focusing on judgment or interpretation rather than *experience* and in misunderstanding the role of felt-bodily sensibility in the name of the (Kantian) alibi of "disinterested pleasure", classic aesthetics is governed by the same estrangement *from* and *of* nature that we find in the hard sciences and in the Enlightenment apology of the (alleged) autonomy of the subject.

First of all, it is not up to the hard sciences to tell us (reductively) what (the experience of) nature is. Rather, the point is to recognize it as the unfailing partner of every emotional and sensory performance of the human being (Böhme, 1995, p. 180), that is, as the element through whose integration the human being also achieves self-understanding. Secondly, two things should be brought to the fore. First, one should highlight the intentionally "naive" criterion of affectivity – of how one feels when experiencing the co-presence of I (or, better, *me*, as already mentioned) and thing (or rather, quasi-thing and situation) – as if it were a very special declination of reception aesthetics. Then, as if it were a very special declination of production aesthetics, one should highlight the competence of the aesthetic work that, objectifying or (*à la* Baumgarten) perfecting sensible knowledge, has specialized indeed in the generation (or at least, evocation) of atmospheres.

Also, my pathic aesthetics rejects the lethal embrace of hermeneutics and semiotics: for these orientations, to put it with the *necessary* and extra-phylological coarseness, the sense of experience is invariably only the object of interpretation-decoding, inasmuch as it is in any case deferred with respect to presence. Instead, a pathic aesthetics aims to present itself as a general theory of (sensible) perception. Of course, this perception is neither experimental nor simply constative, but lived and affective (cf. Griffero, 2014a, pp. 11–59). Precisely because of the simultaneous felt-bodily presence of subject and object,[26] this perception necessarily takes the shape of a (pre-reflective and ante-predicative) affective involvement created by impressions radiated by spaces, things and quasi-things. The sense of the latter, incidentally, coincides in all respects with their felt-bodily appearance *in act* (active and effective, indeed *wirklich*). Take a bad smell, for example: not only do we perceive it, but we also really feel sick because of it (cf. Böhme, 1989, p. 47), regardless of any exact measurement of the intensity of the smell. It is precisely in this sense that, as human beings, we always have aesthesiological experiences which, through both the different senses and the polymodal (or synaesthetic) sphere, turn out to be paradigmatic for a better understanding of our more general affective situation (*Befindlichkeit*).

As a general theory of felt-bodily perception interested in "man as a natural being, sensibly and emotionally affected by his surrounding world" (Böhme, 1989, p. 45), a pathic aesthetics largely coincides with a more general phenomenological theory of atmospheres. As we shall see several times, atmospheres are felt-bodily resonances that, being aroused in us by our (lived, experienced) spaces, ultimately mark the quality of our entire existence. My assumption here is openly anti-psychologistic: in their prototypical form, atmospheres are to be understood, in fact, as feelings that are not internal to the subject, but poured out into one's peri-corporeal and external space. My pathic aesthetics rejects a rigid mind/body dualism as well as the oblivion of affective spatiality, both due to the introjectivist paradigm for which feelings can only be internal.[27] The latter, in fact, finds it almost impossible to explain how one can come out of such internal world to acquire some reliable knowledge of the external one. Rather, my pathic aesthetics wishes to highlight – also from an aesthesiological standpoint – the (felt-bodily, pre-dimensional and pre-geometrical) lived space in which atmospheres hover with their quasi-thingly nature.[28] Physically ineffable, though sometimes actually condensed in a certain area or anchored to some object that may be their primary cause, they still appear "external" enough to suddenly attack us, take over and then, just as suddenly, abandon us. Moreover, they appear so external – almost as objective as streets or at least as clouds[29] – that they are impervious to the projective-interpretive impulse which, in modernity, was ascribed an absurd omnipotence due to the centuries-old but erroneous belief in the semantic-affective emptiness of the external world.

Despite the inhibition sometimes opposed to them by what Schmitz calls "personal emancipation", atmospheres are so influential and authoritative as to arouse total mimic precision, even on the largely unconscious level of so-called virtual movements. Due to their unavoidable vagueness and twilightness, atmospheres are foreign to any reductionist-physicalist approach and, despite the needs of any ontological deflationism, they cannot be ignored. Indeed, atmospheres are quasi-objective, as shown by the felt-bodily and intersubjectively verifiable resonance – because of their quasi-thingly intrusiveness and aggressiveness[30] – that they inspire in the percipients, at least within culturally and historically homogeneous circles. It would be difficult to underline the pre-linguistic and non-physicalistic intersubjectivity or the quasi-thingly and ecstatic quality of atmospheres better than Tetsuro Watsuji. The somewhat Heideggerian Japanese thinker of the first half of the 20th century describes atmospheres with wonderful grace and exquisite naivete:

> One morning we may find ourselves "in a revived mood". This is interpreted in terms of specific temperature and humidity conditions influencing us externally and inducing internally a revived mental condition. But the facts are quite different, for what we have here is not a mental state but the freshness of the external atmosphere. But the object that is understood in terms of the temperature and the humidity of the atmosphere has not the slightest similarity with the freshness itself. This freshness is a state; it appertains to the atmosphere but it is neither the atmosphere itself nor a property of the atmosphere. It is not that we have certain states imposed on us by the atmosphere; the fact that the atmosphere possesses a state of freshness is that we ourselves feel revived. We discover ourselves, that is, in the atmosphere. But the freshness of the atmosphere is not that of a mental state, as is shown best by the fact that the morning feeling of freshness is embodied and expressed directly in our mutual greetings. We comprehend ourselves in this freshness of the atmosphere, for what is fresh is not our own mental state but the atmosphere itself. So we do not need to go through the process of examining

> others' mental states to be able to greet each other with "Isn't it a lovely morning?". Such climatic burdens or impositions occur very frequently in our life.
>
> (Watsuji, 1961, pp. 14–15)

But how should an atmospheric aesthetics be developed? A first possibility is to rethink in atmospheric terms not only the general palette of complex feelings (abjection, serenity, solemnity, oppression, power, sacredness, equivocality, fear, etc.) but also the main notions of classic aesthetics (beauty, grace, sublime, picturesque),[31] up to recognizing some subtle atmosphericness, perhaps dull and anonymous or simply below the threshold necessary for its manifestation and conscious reception, even in situations that normally seem devoid of it. On the other hand, of all possible atmospheres, traditional aesthetics has obsessively examined only beauty, produced by the so-called fine arts, while also mitigating its engaging force.[32] The atmospherologic approach seeks to undermine this hegemony of beauty, recognizing instead that beauty is "the atmosphere of an evident presence that attracts us and increases our vital feeling", but also reminding us that "art is only a particular way of dealing with atmospheres". In other words, what is needed is a more general "politics of atmospheres" aimed at affecting and enchanting the percipients (Böhme, 2006, pp. 20, 26). But it must also be noted that artistic atmospheres are not strictly what they refer to and, paradoxically, not even what they literally represent: rather, they are what "is really present with and in them" (Böhme, 1989, p. 152) and finds a faithful sounding board in the percipient. The categories of traditional aesthetics thus finally lose their asphyxiating "differentiating" nature, ceasing to be (as Kant wanted) a simple subjective feeling – though elevated by the qualification of "transcendental" – and return to being expressive qualities rooted in reality itself and autonomously radiated by it.

The specific felt-bodily way in which something (a thing or quasi-thing, a person or an environment, etc.) expresses its sensible presence by physiognomically condensing the potential of impression, be it synthetic or dystonic with respect to our previous mood: this is how atmospheres underlie our entire pathic life. Indeed, our pathic life is always attuned differently, "inviting" us not so much to do something but rather to "feel" something. Atmospheres are the peak of our perceptual experience of the surrounding world (especially the involuntary experience of it): as such, they are the condition of possibility of a pre-dualistic experience, as in them one cannot differentiate between subjective and objective (we shall come back to this). This is why an atmospherological aesthetics focuses not only on art, but also on the expressive qualities immanent in more impalpable natural phenomena, such as twilight, brightness, darkness, the seasons, the wind, the weather, the hours of the day, the fog, the greatness and power of nature – the sublime, if you wish, but without its Kantian intellectual-anthropocentric interpretation. This is why an atmospherological aesthetics also focuses on relatively artificial phenomena such as, for example, a townscape, music (above all as an art space, or soundscape), natural and/or ecclesiastical sacredness, dwelling and *genius loci*, nuances in art and photography, the charisma of a leader, and so on.

But I wish to immediately dispel the suspicion, already mentioned in the Preface, that this pathic aesthetics may promote regressive sentimentality. In the framework of the aestheticization of daily life imposed by our late-capitalist society, this aesthetics does not seek to a-critically enhance the mise-en-scène or the languid abandonment to idyllic landscapes, but rather to promote a real return to *materiality*. Paradoxically, today's culture of mise-en-scène is what fosters the atmospheric power of materials (new and old),[33] whether they are used in original and effective marketing solutions or in artworks whose materiality, escaping the traditional form–content dualism, aims almost exclusively at generating an atmospherically attuned space. Lastly, it is worth recalling the fundamental importance of architecture in this sense, because "in feeling ourselves we somehow feel the place in which we find ourselves. We experience our own presence and at the same time we feel the space in which we are present" (Böhme, 2006, p. 110). Precisely because it is founded on feeling-oneself – which "means, on the one hand, finding oneself in a space and, on the other, feeling like this or that, being emotionally tuned in this or that way" (ibid., p. 122) – architecture, as the art of experienced and affective space,[34] should be probably placed at the core of a pathic and atmospherologic aesthetics.

The criticism entailed by *this kind* of aesthetics against both ontology and ethics is significant. The insistence on the reality – that is, on the activity and efficacy – of appearance involves two things. On the one hand, the questioning of the traditional paradigms of the reistic model and of the dyad formed by the substance (supposedly essential) and by its properties (supposedly accidental). On the other hand, the full ontological legitimization of the innumerable expressive qualities (characters, physiognomic traits, atmospheres, etc.) suggested by the external world: such qualities are active even if ephemeral, elusive, not strictly describable and sometimes even devoid of a real substratum. This is why atmospheres, as quasi-things, play an ontologically subversive role: first of all because they are *Gestalten* whose phenomenal appearance does not at all refer, as claimed by traditional physiognomics, to some internal dimension unjustifiably considered more true and essential. Secondly, because the involving action of atmospheres, completely coinciding with the resonance aroused in the felt body of the percipient, does not at all need to be traced back to some previous and autonomous cause: the latter is only needed within a thought aimed at prognosis and therefore also at manipulation. Thirdly, because atmospheres constitute a sort of in-between prior to subject and object – in other words, a relationship that precedes the things related (a true scandal for logic!),[35] which only become such (namely two autonomous and discrete poles) *après-coup*. Finally, because, despite the ephemeral character and the temporal intermittency that characterizes their manifestation (and, in this perspective, *ipso facto* their existence), atmospheres worthy of this name are always powerful and therefore exert a veritable authority (cf. Griffero, 2014c).

So, for a pathic aesthetics it is fundamental to claim the importance of appearance and ephemerality: dimensions to which Western thought and common sense refuse to ascribe a true reality, preferring to bet everything on full things, endowed with borders, separated from others, perduring in time, and normally inactive if not touched. But I wish to make it clear that this apology of appearance does not transform *ipso facto* a pathic aesthetics in an ethically indifferent variant of the postmodern "anything goes". On the contrary, its ambition is to tackle today's "aesthetic economy" without nostalgic temptations and to encourage, if not a new "aesthetic education of the human race", at least the "sensitive and felt-bodily experiential potentialities of the human being" (Böhme, 1989, p. 15), of which one becomes aware thanks to art but also thanks to the daily "ecological" perception, above all if oriented towards (atmospheric) appearances generated precisely by the so-called aesthetic work. Of course, Schiller's utopian and conciliatory "play" is thus replaced by the atmosphere, which is precisely "something mediating, something that mediates, [which] is an in-between, between subject and object, between defining and receiving, between doing and feeling" (Böhme, 2006, p. 52). Being critically educated in this in-between (*Zwischen*) – which *qua* atmosphere can only be separated *ex post* from the two poles involved[36] – means learning how to "treat" atmospheres with the necessary competence on the operational-creative and receptive-perceptive level (which may also mean knowing how to occasionally dissociate from them). It is only because of this acquisition of an atmospheric awareness that one can – of course, only within certain limits – escape the most alienating and manipulative outcomes of today's aestheticization.

A pathic aesthetics is strictly focused on appearance as such: that is, more on the felt-bodily resonance of appearances than on their material supports (which sometimes are even absent) or on their semantic values (intended for a more intellectualistic approach than the one promoted here). However, the pathic (neo-phenomenological, atmospherological) aesthetics that I am outlining does not at all intend to silence its critical potential. In fact, only an expert in appearance can, for example, deactivate the mise-en-scene that is passed off as a reality *tout court*, developing the necessary critical competence "of usurpation, of manipulation and the suggestion exercised by the production of atmospheres upon those who are exposed to them" (Böhme, 2017b, p. 34). Now, Schmitz (1998, p. 182) rules out the intentional generability of atmospheres, degrading it to a simple "technique of impression" (in the pejorative sense). Instead, while de-axiologizing the concept of "atmosphere" and admitting that it's not like *any* situation can generate *any* atmosphere,[37] I think that any spatially diffuse affectivity is an atmosphere. And this is so regardless of whether it was produced involuntarily or whether its agency is not even human (climate, natural phenomena, etc.), or whether it was produced by specialists for the purpose of obtaining something, be it commercial success or political consensus.

This atmospherological aesthetics undoubtedly falls under *realism* (a term that should possibly be preserved both from the media abuse it has been receiving and from its dogmatic-conservative implications): in fact, with respect to atmospheric qualities, it rejects the lazy explanatory hypothesis offered by associationism and projectivism. It is, I believe, a theory that values in a relatively new way the cooperative relationship between human perception and the ecstatic nature of the world:[38] that is, a pathic intermediate (in-between) but pre-dualistic area which is still to receive a proper philosophical and aesthetic analysis. As mentioned, the keystone of this "in-between" is always the felt body (*Leib*), that is, bodiliness *qua* experienced (and not in the physiological-anatomical sense). Far from being an unmodifiable "fact", it rather appears as a "task", as something we are daily responsible for with respect to the quality of our entire existential condition, which is increasingly subject to (perhaps threatened by?) technological changes and prostheses.[39] A pathic aesthetics thus rethinks human felt-bodily life, as well as the experience in which, feeling ourselves, we also sensibly perceive the world – where, naturally, saying that only sentimental experiences are lived atmospheric feelings is an obvious psychological statement but certainly not an explanation of their provenance or their ontological nature.

This reconsideration, in my view, is equally important from the aesthetic and from the ethical point of view. On the other hand, the failure of the perceptive illusion in the artistic sphere and of the normative-argumentative illusion in the ethical field is probably due to two exquisitely aesthesiological paradigms: that of a certainty conceived as an unregulated *experientia vaga*, and that of evidence infallibly demonstrated by ubiquitous and yet unrecognized involuntary vital experiences, too often degraded to insignificant automatisms. Far from promoting ideals of ataraxia and autonomy, a pathic aesthetics rather promotes the (never fully uncritical) acceptance of our affective hetero-determination, that is, of something that certifies better than other traditionally privileged states (including the *cogito*) our being-in-the-world. Indeed, the immersive exposition to a particular "aesthetic" presence, to a perception that is somehow "extorted" from the outside – and within which the phenomenological *that* turns out to be irreducible to the cognitive *what* – expresses our lives better than anything else. Because our lives are first and foremost pathic.

A pathic aesthetics, which is atmospherological (as we have seen) and neo-phenomenological (as we will see later), therefore is not a reflective theory on privileged objects, but rather a practice through which to learn "how to expose ourselves" (Böhme, 1994, p. 189) to what happens, in an experience that treasures whatever we may unexpectedly encounter. This is a pathic orientation precisely because it recognizes the primacy of feelings, which atmospherically grasp us from the outside and involve us in a felt-bodily way even beyond and against our will. This *pathos* – which, as said, has nothing to do with *dolorismo*,[40] and does not consider vitalistic efficiency a privileged way to have a rich atmospheric phenomenal experience[41] – can be understood much better in ethical-aesthetic terms than in terms of action. And this is the reason why, rather than the subject – pathologically overestimated by modernity with the well-known negative consequences – we should here rather focus on the pathic (or perceptological) *me* that precedes the egological solidification.

The general philosophical value of this (here only sketched) project of a pathic aesthetics, of a deliberately oxymoronic (non-sensistic) thought of the senses,[42] should be sufficiently clear now. It aims at a non-gnostic ("non-expert", to use a different vocabulary) but affective-emotional investigation of our felt-bodily involvement in atmospheric feelings effused into (lived) space, which are as independent from us as (Husserlian) passive syntheses. This truly interdisciplinary *atmospheric turn* has also reignited topics that had long been mistakenly confined in that junction between the 18th and 19th century which was rightly called (somewhat mockingly) "age of sentimentalism (*Empfindsamkeit*)". In light of this, it is finally possible to conceive of humans – without regressive (or frankly reactionary) temptations – *not as "subjects of" but rather as "subject to"*. We are sovereign and adult (rather than independent and autonomous, as claimed by modernity) not when we avoid being exposed to what happens to us, but rather when we learn how to do so in the best way (cf. especially Böhme, 2008, pp. 188–201). And we are sovereign or self-confident with no need to neurotically repress the negative or what happens to us, which hetero-determines us much more when we fight it than when we abandon ourselves to it (*cum grano salis*). This is the goal of *the present* ambitious pathic aesthetics, for which everything that acts is real, insofar as it is experienced (Klages): to learn how to live with the affective involvement we are subject to with the affective immediacy and necessity which our paranoid culture inevitably sees as occult and alienating, and which instead constitute an emancipatory opportunity.

Notes

1 For more on this drastic downsizing cf. Schmitz (1977, pp. 613–685; 1990, pp. 455–497; 2007, pp. 81–93).
2 Something obviously unheard of for Hegel, who "knows nothing of the pathic moment of sensation, of its communicative dimension" (Maldiney, 1966, p. 220).
3 Which is interested not so much in imitating feelings that are external to the work but rather in releasing them from the work itself: with Bacon, for example, one could speak not of the depiction of affects but of a representation steeped in affectivity (Busch, 2007, p. 63). This is nothing new, however: just think of Artaud's anti-cathartic view of theatre as a "plague" and a subcutaneous (psycho-physical) infection.
4 Developed by Hermann Schmitz starting from the first volume of his System (1964). For a first introduction cf. Schmitz (2019) and Griffero (2019).
5 This escape can also take place in the metric space, which demonstrates better than anything else, I think, the quasi-objectivity of atmospheres. For an overall

reconstruction (in the light of neo-phenomenology) of the problem of atmospheres cf. at least (sticking to the most programmatic and philosophical titles): Tellenbach (1968), Schmitz (since 1969, up to the latest 2014), Böhme (1989, 1995, 1998, 2001, 2006, 2017a, 2017b), Hauskeller (1995), Hasse (2005, 2012), Griffero (2014a, 2016a, 2017a), Andermann and Eberlein (2011), Rauh (2012, 2018), Bulka (2015). For an initial response to the objections to atmospherology cf. Griffero (2017a, pp. 19–53). For a more complete list of bibliographical references, which are constantly increasing, cf. https://atmosphericspaces.wordpress.com/literature/.

6 In my opinion, therefore, atmospheres do not "almost always" pervade situations, somehow "hanging" on them (Schmitz, 1998, p. 179; 2003, p. 248; 2014, p. 60). Indeed, it is precisely the affective and felt-bodily atmospheric tone (not necessarily apparent) that makes a situation stand out and thus distinguish itself from others. Of course, the changeable situation of the subjects involved can induce them to (geographically, historically, culturally) accentuate even "relatively" different atmospheres, and a situation can gradually de-emotionalize the atmosphere it generates and/or welcomes (cf. Grossheim, Kluck and Nörenberg, 2014, p. 64). However, in this case the outcome would simply be a "colder" or even dull and anonymous atmosphere – certainly not a situation that is fully devoid of atmosphericness.

7 Here I am not referring to Schmitz's distinction between linguistic arts, which clarify ambiguous situations in a necessarily economical form, and non-linguistic arts, which do not so much express the sentiment but rather "present" sentimentally saturated impressions as such.

8 What acts as a sounding board, and therefore as the atmospheric-marker, one could say, of spatial feelings, is precisely the felt or lived body (*Leib*) (see Griffero, 2017a, pp. 55–67). The latter is composed not by organs but by "felt-bodily isles" of variable duration and configuration; it cannot be reduced to the skin border and only occasionally coincides with the physical body (and with the so-called body scheme).

9 As in the fascinating case of vicarious shame (Griffero, 2017a, pp. 79–92).

10 Here I follow the dialectic proposed in many neo-phenomenological texts (starting from Schmitz, 1980) between personal regression and personal emancipation (cf. Chapter 5).

11 Hermann Schmitz rightly distrusts the idea of tracing the authority of evidence back only to coherentist-consensual principles such as rational dialogue (*à la* Habermas). He denounces such principles' relapse into trials condemned to regress *ad infinitum* or to feed on an optimistic-eschatological illusion (immediately disproved by facts). Also, they have to be content with the (surely unquestionable) procedural value of the best argument or with controversial pragmatic principles such as utility in view of subsequent experiences. On the other hand, consider the curious attempt (Lauterbach, 2014) to establish a dialogue, in terms of philosophy of religion, between Habermas and Schmitz (a dialogue that never took place even though they both studied in Bonn under the guidance of Erich Rothacker). This dialogue could be based on the idea that the core of religious experience, declared impenetrable by the agnostic Habermas, may instead become accessible thanks to Schmitz's neo-phenomenological approach (erroneously defined romantic by Lauterbach, 2014, p. 33) to subjectivity. In particular, it can become accessible precisely thanks to the concept (fundamental here) of absolutely binding feelings, which in some cases – as in the "archaic shame" felt by the survivors of a catastrophe, for example, or in the melancholic atmosphere conveyed by a priceless sacrifice – are even admitted by an illuminist like Habermas (cf. Lauterbach, 2014, pp. 560–565).

12 An evidence that, in turn, is perhaps even more affective than Schmitz himself admits: cf. Werhahn (2011, p. 58). On the other hand, a person could identify with the "ought to be" (*Sollen*) to the point of seeing it more as a fact than as a hope.

13 At the basis of law there would therefore be not belief or will, but "a juridical presentiment which takes the form of abstention, a sensibility for which certain forms of life and certain behaviors, when bearing their extrinsecation with caution, can sufficiently resist anger and shame while others appear as intolerable provocations of anger and shame" (Schmitz, 1990, p. 388).
14 "The human being cannot live except under norms, because they live in situations containing programs that promote their obedience": a norm is in fact "a program for possible obedience" (Schmitz, 2012, pp. 7, 14).
15 Schmitz in Werhahn (2011, p. 130).
16 In fact, one can legitimately feel shame even in the face of the *nomos* of one's own community, and one can feel anger even for the offense against an impersonal value such as music, for example when hearing someone butchering a wonderful piece.
17 The Christian solution is notoriously different (Jn 8:7, "let him first cast a stone"), since it turns anger towards the guilty, which would demand revenge, into collective shame (Schmitz, 2003, p. 302).
18 For, in itself, revenge would only produce new anger, and shame would petrify those who feel it, thus preventing its cathartic outburst.
19 It is no coincidence that in non-juridically regulated cultures there is a tendency to respect anger (felt by the harmed person), which is evidently believed to be endowed with exceptional and legitimate authority.
20 But forgiveness is also cathartic in this perspective, since it takes seriously both the guilty and the damage suffered.
21 As Schmitz (2012, p. 189) seems to think. For him, an aesthetic argument could still end with an inoffensive shrug.
22 Here I freely apply some terms of the felt-bodily dynamic studied by Schmitz.
23 Somewhat inexplicably, Schmitz circumscribes beauty only to what it is before us (Schmitz, 1990, p. 494), excluding its potential motor and ambulatory genesis.
24 A poem, for example, induces "a commotion, a chance of loss, as if the ice on which the wise adult leads their life became friable for a moment" (Schmitz, 1990, p. 467).
25 For its structure, see Böhme (2001, 2006).
26 "The perceiving subject is real in his participation in the presence of things, the perceived object is real in the percipient presence of the subject" (Böhme, 1995, p. 177).
27 "For every individual endowed with conscience, the world is split between their own external and internal worlds, with the proviso that they will become maximally aware of an object of their own external world only insofar as such object has a proper representation within the internal world of the individual" (Schmitz, 2007, p. 14).
28 I have tried to elaborate this notion by Schmitz elsewhere (cf. Griffero, 2017a).
29 Schmitz (2014, p. 9) believes it necessary to rectify this in relation to his previous comparison between atmospheres and streets (Schmitz, 1969, p. 87), considered too reifying and dictated by the anti-introjective polemic. From a generally anti-reifying viewpoint, I don't think this rectification is strictly necessary, since there are much more solid and thingly entities than atmospheres which, in the light of a slightly more heterodox ontological investigation, have boundaries that are as indeterminate and incomplete as the atmospheres they may radiate (Griffero, 2014a, p. 66; 2017a, p. 26).
30 On the ontological characters of quasi-things (emotional invasiveness, apprehensional non-perspectiveness, a-historicity, pure actuality, encorporating and/or excorporating communicability, intermittent existence, bipolar causality, predimensional spatiality, para-fractality) see Griffero (2017a, pp. 9–14).

31 This is the case regardless of the choice, which I find very problematic, of seeing the aura as "the atmosphere lacking in characteristics, that is, the atmosphere in general" (Böhme, 2006, p. 15).

32 This is evidenced by the absence of a specific expression to indicate the reaction to artistic beauty (Böhme, 2010, p. 177): a serene landscape makes you serene, but a beautiful artwork does not make you beautiful – at most it increases your vital feeling (as Kant posited).

33 "We feel materials to the extent that the atmosphere radiated by them enters our affective situation" (Böhme, 1995, p. 54).

34 Cf. Böhme (2006, 2017b) and, also for the literature mentioned in it, Griffero (2014d).

35 This scandal is not such for phenomenology, though: "The present being of sensible feeling – and consequently sensible feeling in general – is the vital experience of a being-with that develops in the direction of the subject and the object" (Straus, 1963, p. 372).

36 "A captivating atmosphere is not a relationship between a subject and something by which the latter is absorbed, but it is inseparably connected to this subject. When, for example, someone gets angry, anger is not a medium for them, as if it were a second thing (a relationship) through which they relate to a third thing (the object of anger with a sphere of condensation and an anchoring point). Rather, the subject is so captivated by anger that they cannot distinguish it from themselves, as if them and the anger were one thing (or two or three things within a numeric manifoldness). Only when they have 'come out' of the anger, to the point of being able to reflect on it assuming a personal stance, can they conceive of anger as something 'between' themselves and the object" (Schmitz, personal communication, 03/12/2014).

37 For example, it is impossible to feel the atmosphere of Christmas at other times of the year, or to feel the familiar atmosphere of a dinner with friends in an anonymous and crowded public place.

38 "The subject's receptiveness corresponds to nature's showing-itself, a coming-out-of-itself of the things of nature" (Böhme, 1995, p. 157).

39 According to Böhme (2008, p. 67), "from our felt-bodiliness come moral problems; that is, serious problems, and as we make our decisions based on those, we thereby decide what we are and how we are as human beings."

40 For an analysis of the atmosphere of pain (necessary for subjective self-awareness) cf. Griffero (2017a, pp. 69–77).

41 While avoiding the existentialist predilection for depressed emotional tonalities and defending the reliability of what acts as support in reality, Bollnow (1956, p. 118) exaggerates, however, in his belief that only when the man "is calm and serene in himself can he easily accommodate also external things as they are".

42 This is how Falter (2003, p. 50) defines Klages' synaesthetic and largely anti-conceptual orientation.

2 The psychologistic-introjectivistic-reductionist paradigm

The source of all errors

"No improvement is too small or trivial to be worthwhile [...] One should never begrudge deletions [...] Nothing should be thought worthy to exist simply because it exists, has been written down" (Adorno, 2005, p. 51). If this warning applied not only to writing, as in this case, but also to the history of ideas, we should consider it entirely legitimate not only to "rearrange" Western philosophical thought, reducing it to a few essential patterns to avoid its disconcerting complexity and chaoticness,[1] but also to point out its mistakes and impasses. Especially in modern times, philosophers have indeed tended to justify their own perspective, claiming to be the first to notice some fundamental error (usually located, if not at the beginning of a culture, at least in rather ancient times, so as to have affected all that followed). Such philosophers have also tended to claim they were the only ones able to interrogate the witnesses of this culture and to have the indispensable tools to get around that mistake, or at least move away from it. If we uncritically accepted this attitude, mis-applying Adorno's warning, we would be spoiled for choice in the Western collection of profound untruths or semitruths (Sloterdijk, 2009, p. 102, fn. 2): from the dogmatism of Scholasticism (Descartes) to the "Ptolemaic" *qua* pre-transcendental perspective (Kant), from the oblivion of Being (Heidegger) to the victory of the spirit on the soul (Klages), from the Platonic-Christian assassination of the tragic (Nietzsche) to the reductionist quantification of the plena (Husserl), from the destructive logic of identity (Adorno) to the masked anthropocentrism (Feuerbach) and to the underestimation of the structure in favour of superstructures that the false conscience considers autonomous (Marx). And so on.

Explaining the history of thought as a consequence of an error is perhaps only a form of paranoia, corroborated by a self-propelling tendency to self-feed even through contrary evidence, thus turning interpretative and demystifying enlightenment into something no less delusional than the mistake one wishes to denounce.[2] In any case, it is fundamental to understand what it means to identify and stigmatize an error. First of all, because to identify an error, it is necessary to presume the possession of some – at least local – truth. And then because, in order to identify an error in the past, one ought to believe to be on a more advanced level of cognition, if only in order to go back to a truth prior to the error, and to which it would be possible to return. It is usually left unexplained how to do so (especially within a linear conception of history) if not by generically presenting oneself as *laudatores temporis acti*. To identify an error so pervasive as to determine the whole history and the very mindset that guides us (or, less strongly, in which we "dwell"), we should also be able to consider ourselves free from such influence. While revoking the irreversibility of time, one should be able to bracket the very *forma mentis* which, although conditioned by the error, is at the same time necessarily the condition of possibility of the discovery of the error. Hence the problem: how and why would an erroneous and pervasively influential metaphysics provide its members, or even just an elite, with the ability to unveil its erroneousness? And why would a philosopher – almost a new version of the transcendental physician *à la* Novalis – be able to diagnose the disease that afflicts the history of civilization and prescribe the right therapy, although often circumscribed, of course, to the semi-prophetic level of ought-to-be?

Then there is another difficulty, concerning the analogies or differences between the history of philosophy and that of science. Explaining the history of thought, or in any case vast sections of it, as errors presupposes that, both in philosophy and in the empirical sciences, there has been progress and that, unlike what happens in a common law system, there are no binding precedents. In other words, there are certain assumptions at play here, which I will now list. a) Far from considering ourselves inferior to the great philosophers of the past, we can safely assert we have refuted and overcome, say, Aristotle; today we cannot see ourselves as colleagues of those who believed in the unmoved mover, just as an astrophysicist would not see a supporter of Ptolemy as a colleague. Therefore, b) we admit that, also in philosophy, knowledge is cumulative, so that it only makes sense to discuss the most recent theories and automatically consider older ones inadequate. Finally, c) we should deal with the problems of the philosophy of the past synchronically, without leaving them to doxographers, but evaluating frankly and with deliberate anachronism the insufficiency (or at least relativity) of those views, which have now been overcome.

These seem to be improbable arguments, and in any case not worthy of philosophy, if the latter is understood not as the epistemological or linguistic reflection on the natural sciences – which are subject to the progress of its themes – but as a conceptual *a priori* investigation of essentially timeless dilemmas and "essentially" unresolvable problems. Thus understood, philosophy, as a natural genus, is in principle as foreign to the accumulation of empirical notions as art: a musician, for example, would not say that they are better than Mozart, but at most would offer a different solution to the "same" problems. And yet why would people still deal with philosophy after Plato, if there wasn't the possibility of some kind of progress? Perhaps one could posit the idea of local progress so that, within a given "paradigm", it is perfectly legitimate to claim that one has overcome the mistakes of a previous (but how far back in time?) philosopher. This solution may appear good, but it is very partial, as it doesn't at all apply to the far-from-local (and therefore more suggestive) reconstructions of the history of metaphysics as a mistake. These reconstructions, oblivious to the fact that any rectification implies both gains and losses, almost always justify their appeal to "go back" to some previous perspective as if it coincided with unquestionable progress.

Instead of further complicating the matter, here I will simply present a relatively recent philosophical perspective such as New Phenomenology, elaborated by Hermann Schmitz since the 1960s, and built in its *pars destruens* on the identification of errors committed in the distant past. I believe that this perspective can indeed play an essential role in my pathic aesthetics. New Phenomenology is not archaeological, in that it doesn't only interpret past philosophers to present their theories as less vain; nor is it anachronistic, in that it is foreign to philological scrupulousness and historical contextualization. However, the neo-phenomenological approach lacks the (unattainable) requirement of a historical-contextualist reconstruction by which, briefly put, one should not describe the view of a past author in the light of criteria they did not and could not have. Indeed, it is perfectly legitimate to present Columbus as the (unwilling) discoverer of America and not of Cathay. Likewise, for a rational reconstruction like the neo-phenomenological one,[3] it is legitimate to present Plato not in his terms, but as the systematizer of psychosomatic dualism (even though it would be absurd to claim he "intended" to do so!), thereby discussing both his problems and solutions, and the epochal (also historical and political) effects of his ideas. Rejecting the holistic causal link coming from the most organicistic historicism, according to which a type of thought would belong exclusively to a certain epoch, one ought to insert an author of the past into a history of effects that exceeds the vocabulary they used,[4] even putting in dialogue thinkers from different times on "common" topics (which appear such *ex post*).

This approach, of course, comes with difficulties. For example, due to the *post hoc propter hoc* fallacy, this view assumes that present events are indirectly (and destinally) due to some previous philosophical reflection (which perhaps was only rediscovered today). Also, it is victim to the precursor bias, by which a past author necessarily held an "implicit" view of some topic which for us is extremely relevant. This leads one to wonder whether history is therefore always written by the winners – a hyperbolic doubt that challenges any *Geistesgeschichte* (at least since Hegel), whether it is ascending or descending. Yet I do not think that such difficulties truly repress the human want for narratives, which amounts to two insuppressible needs. The first is that of feeling like a member of a community that also includes those who have died. The second is that of being able to count on a criterion of orientation in life (even in its historical course), which, by suggesting identification with or rejection of a tradition, compensates for the frequent disorienting effect of historical distance, especially due to the irresistible acceleration typical of modernity. And the question, more theoretical than historiographical, is the following: why is it that New Phenomenology is not content with posing different, and perhaps better questions, than its predecessors, and claims instead to give better answers to the same questions, evidently judged fundamental and eternal?

Schmitz denies conflicting with the history of philosophy, at least – as he says without a hint of irony – with philosophy up to the so-called Presocratics (!). As for the subsequent history (and philosophy), dominated by abstraction and above all by the dogmatization of fictions, acceptable only as long as one thinks of them as such, he clearly is in conflict with it. Let us put aside here Schmitz's controversial historical-political aims,[5] as well as his polemical thesis that nothing truly new emerged between ancient Neoplatonism and Kant. Instead, let's concentrate on Schmitz's analysis – which I have partly already borrowed to give an adequate context to my pathic aesthetic – of the source of all errors: i.e. the psychologistic-reductionist-introjective paradigm and its nefarious consequences (dynamism, autism, ironism). Starting from the very beginning, one might say that, by focusing on the felt body rather than on the incorporeal self-reference dear to existentialism, New Phenomenology redefines philosophizing as a reflection on what one feels with the felt body in the space in which one dwells (Schmitz, 1964, pp. 14–27).

What follows is the need for a description that, rejecting any constructivist implication, conflicts with the kind of selective abstraction (Schmitz, 2005b, pp. 19–20) – in a nutshell: Indo-Germanic syntax, post-Democritean metaphysics, ontology based on things, properties and relationships – that in Western culture, obsessed with the reductionist idea of tracing complex experiences to some "nothing but",[6] leads to neglecting sensible (phenomenal and sentimental) life. Instead, New Phenomenology looks to the "most delightful and ideal state of nature" thanks to which "the songs of Homer have the power to deliver us, if only for brief moments, from the fearsome load with which tradition has weighed us down over many thousands of years" (Goethe, 1998, n. 662). In this way, in antithesis to modern rationalism, this approach strives to bring back to the fore an archaic paradigm that, after the 5th century BC, only survived in Neoplatonism (and paradoxically despite Plato)[7] or in minority and heterodox enclaves (Pythagorean numerology, humoralistic medicine, astrology, mysticism, natural magic, paracelsianism, alchemy, popular philosophy, etc.). This paradigm is opposed to reductionism, thus making available "a basis of abstraction that is closer to life experience compared to what has governed the prevailing intellectual culture in Europe ever since Democritus, Plato and Aristotle" (Schmitz, 2007, p. 15).

Also, this paradigm reacts to the above-mentioned "fearsome load", characterized in the Western tradition also by the unbridgeable split between an "affirmative culture", considered more legitimate and binding, and a "culture of fantasy", which instead comprises the whole vast sphere of vital experience, obliterated by the dominant paradigm, which only attributes it a decorative-compensatory value. This tragic and epochal drift, which can be placed between 450 and 350 BC, is therefore the cause of the errors underlying most of the history of the Western spirit, some of whose consequences are under everyone's eyes: the weakening of superindividual spheres (the cultures of law and religion, for example) and collective situations (rooted in a *nomos*),[8] followed by an autistic-individual fragmentation and an irresponsible ironic scepticism; or the acceptance of the dogma of immanence, followed by solipsistic hedonism (simplified in terms of pleasure/displeasure) and by the intensification of the processes of self- and world-empowerment (theological but also technological).[9] And these are only some of the many possible examples. In fact, the whole Western culture – including the abstract (statistical-experimental-combinatorial) background of today's physics, still inhibited by the unavoidable delegation to an omnipotent God before its definitive mathematization – originates for Schmitz from a primal mistake committed by the Greeks, after they won the war against the Persians and took the path of science, aesthetics and democracy (cf. Schmitz, 1999a, p. 381). And the first one to "blame" is Democritus. He is responsible for replacing dynamism with kinetics, for intra-human dualism between soul (understood as a material house, for the first time) and the material body (declassed to the latter's tool), and for the reduction of outside expressive qualities to atoms, conceivable as stable invariants in the incessant flow (hence also the contraposition between matter and form) just like Platonic ideas.

This paradigm shift – a necessary change of direction, if we embrace Schmitz's (dualistic and therefore certainly not theoretically neutral) rail junction model – caused the genesis of a thought that has three main features. First, it is combinatory-prognostic because it is based on singularism, i.e. the nominalist dogma that everything is singular. Second, it is constructivist because it rests on projectivism, i.e. the idea that external significance is but the outcome of a human projection (even pragmatically conditioned) on a world made up of mere quantitative elements – a prejudice become virulent in Nietzsche and especially in early 20th-century Nietzscheanism. Finally, it is constellationist, based on understanding the world as a network of individual (singular) factors.[10] Plato himself is allegedly but an epigone of this fatal turn (cf. Schmitz, 2003, pp. 348–363), even if he became a sort of *spiritus rector* (Schmitz, 1990, p. 17) of what followed. Aristotle opposed Plato, but nevertheless was responsible for other serious mistakes, such as the elaboration of a tripartite ontology (things-properties-relations) as opposed to Platonic ideas. The problem with Aristotle's ontology is mainly that it disowns all that exists while not being any of the three terms mentioned, and it does not give priority to relations as opposed to the *relata*. This turn was anthropological-pedagogical rather than strictly theoretical, as it was due to the human need for self-and world-empowerment, which characterizes the human being as an autonomous person no longer subject to the arbitrary tyranny of external forces.

Convinced of offering today's humankind a (now) unheard-of opportunity for self-understanding, Schmitz decisively rejects this turn through an appreciation both of the *logos* inherent in the alphabet of the felt body, and of the archaic felt-bodily dynamism explicitly aimed at relativizing apologetic notions (soul, creativity, etc.) which are nothing but phenomenologically unjustified artificial constructs. This dynamism[11] can still be found in the Pythagorean dynamic of polarity; in Empedocles' idea of the conflict between love and anger as spatially extended atmospheres; in Aeschylus' conception of resentment as a superpersonal atmospheric mood; in Heraclitus' metaphorical contraposition between bow and lyre; but also, residually, in the stoic doctrine of the dynamic of the *tonos* (exemplified by the felt body) and even in St Paul's idea of spirit and flesh as conflicting atmospheric forces in the human body.

For Schmitz, who objected to the hypothesis of a Homeric unitary psychology and anthropology, the best example of such "antelapsarian" *forma mentis* lies in the *Iliad* (cf. Schmitz, 1965, pp. 365–445). This (non philological) Homer, for whom humans are exposed to the obsession or affection of gods,[12] sees human life experiences as not purely psychic: they are decentred and exposed to a "concert of semi-autonomous sources of impulse, the most important of which is *thymós*" (Schmitz, 2003, p. 342).[13] Life experiences are not only psychophysically undifferentiated as still related to an image of the world that prefers the distinction between organ and its function over the dualism between body and spirit (or soul); in Homer they are mostly the example of an age that explains everything on the basis of felt-bodily impulses[14] and not based on the differentiation of sense organs. At that time, all that would later be defined "psychic" was taken to be felt-bodily with the same one-sidedness with which modernity would make the body psychic in every respect. Homeric times relied on a felt-bodily motor schema, based on the irreversible directions and on the swaying of diffuse felt-bodily isles[15] and not on the perceptual body schema (the habitual conception of one's own body) that modern psychology would derive from the experiences made through the senses.

Therefore, the presocratic age[16] still used a pluralist bodily vocabulary, which – being the eminent model of a pathic life form – still envisioned human beings as almost completely exposed to (to the point of being possessed by) divine-daemonic external aggressions, to whose uncontrollable power they were subordinate (just as today they are believed to be subordinate to the residual impulses of hunger and thirst). Yet things changed with Socrates and Plato, and indeed already in the *Odyssey*. Here, in the frame of a struggle between *psyché* and *thymós*, the hero (who in this sense is already fully modern)[17] can already distance himself from his outer felt-bodily impulses, exerting complete control over them as well as over his facial expression. In Odysseus, personal emancipation runs now parallel to his rationalism, which declasses the divine-numinous to a *daimon*, reduced to a predictable partner (who, in any case, one can pray to). But the most harmful consequence of this shift is the introjection of feelings into a closed-off inner sphere.[18] The latter then appears to be conceived as a solid body: a metaphorical multistory house (mind, reason, intellect, senses) furnished with all the meaningfulness (intentional acts, ideas, memories, impulses, sensation, feelings, cf. Schmitz, 2003, pp. 333–334) that, before, was mistakenly believed to belong to the external world and is now believed to be governed by reason.

This privatization of the emotional sphere then continued, made easier by the progressive disappearance of transcendent restraints and fears. In the Christian Middle Ages (for Schmitz, 313–1303 BC) it also involved the believers' happiness, as they had to control the drives of a body already reduced to a machine (Augustine). Then it gave rise to a singularist ontology (Ockham) for which reality, made up of singular and discrete entities, lends itself to the most arbitrary constellationist combinations, producing, as its ultimate consequence, technical expertise, the market economy and even today's computer digitization.[19] In parallel, the power of affective involvement, first attributed solely to God, then moved on to Western Christianity, through various compromises and de-mystifications (papacy, the Crusades, imperialism, etc.) aimed at adapting the divine to human needs, even placing the former in the hands of the latter. In either case, their aim was to deprive God of any numinous casualness in favour of fixed (even Trinitarian) properties responding to human needs; in fact, humans can address God through prayer, which is no less than a form of injunction (Schmitz, 1999a, p. 38 ff.). And all of this, for Schmitz, says enough about the – no longer only elitist-aristocratic – strive for self-control and the search for a no longer transcendent, but earthly wellbeing (capitalism).

Notes

1 Almost a conditioned response, in the tradition of thought, ever since the brief and far from objective history of philosophy offered by Aristotle in the first book of *Metaphysics*.

2 Indeed, it would be characterized by "the same quality of religious revelation. And the revealed truth of a religion cannot be changed, because its modification would not be correction but heresy" (Zoia, 2011, p. 31).

3 Here I adopt the definition given by Rorty (1984). Adopting the distinction between meaning and significance (see Griffero, 1984, 1988, p. 182 and passim), which is essential for all recognitive hermeneutics (Emilio Betti, Eric Donald Hirsch jr.), one could also say that a rational reconstruction is only interested in significance.

4 We do not know how much the author can then rightly be said to be irremediably compromised with the basic assumptions of this vocabulary. On the non-identifiability of a concept with its tradition of use cf. Marconi (1988, p. 37).

5 Schmitz's *desiderata* include no less than the reconstruction of a sort of Roman empire, thanks to the valorization, finally made possible by the mediation of reunified Germany, also of the principles of Eastern Christianity, and especially of the idea, traced back to Neoplatonist sources, of *sobornost* or organic solidarity (Schmitz, 1999a, pp. 396–404).

6 To the physical datum in the natural sciences, to language in analytical philosophy and (today) to the brain in neurosciences.

7 Schmitz seems to refer, above all, to the Neoplatonic explanation of perception in sympathetic terms.

8 Plato's Socrates already dissolved the holism of "rooted situations" with typically reductionist-constellationist reasonings.

9 Monotheism itself would be functional to the control of a divine atmosphericness, originally more varied and disturbing (Schmitz, 1977, p. 177).

10 Not coincidentally, these factors have always been exemplified by alphabetic letters, regardless of the fact that, as Aristotle already noted (*De gener. et corr.* 315b, 1416), their outcome may be either a tragedy or a comedy!

11 For a summary, see Schmitz (2019, p. 52 ff.).

12 In this sense, complexes (Freud) and archetypes (Jung) are allegedly similar to Homeric impulses, while lacking a felt-bodily localisation (Schmitz, 1990, p. 200).

13 Note also *kradie* (the heart as the trigger of initiative), *noos* (the trigger of the observational impulse), *noein* (what happens), *ätor* (the trigger of the reactive impulse) and *phrenes* (diaphragmatic), wise and reasonable.

14 Whose local-qualitative multiplicity justified the lexical multiplicity for their pseudo-organs.

15 Cf. Schmitz (2011, pp. 21–23). On felt-bodily isles cf. Griffero (2017a, pp. 55–67).

16 A misleading term, to the extent that it overestimates the figure of Socrates (fictionalized by Plato).

17 Note that this reconstruction (Schmitz, 1965, pp. 445–451; 1969, pp. 413–418) is independent from and yet largely similar to that proposed by Adorno and Horkheimer.

18 For his criticism of introjection, Schmitz refers to Avenarius (2014). But cf. also Haensel (1946, pp. 63, 299): "what the intellect 'projects outwardly' outside itself is only the names and images gathered in its memory, and certainly not the powers that excite the soul"; the error therefore lies in the fact that "the powers that move the world are introjected in the self, they are made its 'internal' powers and are then re-projected onto the world", thus obtaining nothing but "individual hallucinations".

19 This link in the history of ideas is suggestive but, of course, problematic.

3 A history of repressions

So far I have summarized the essential features of what, according to New Phenomenology, is the source of all errors, namely the psychologistic-reductionist-introjectionist paradigm and its success in the history of Western thought. This paradigm has caused many serious repressions, in the Freudian sense of the term. I shall now discuss them, while also listing – *e contrario* – the cornerstones of a possible pathic-neophenomenological aesthetics.

a) The first repression regards the body, degraded from felt or lived body (*Leib*) – whose felt-bodily voluminosity was (and is) experienced independently of its spatial and physic dimension – to a material-anatomic body (*Körper*), understood as spatially and physiologically delimited, perceptible from the outside and arbitrarily manipulable. With the success of the dualism between (unextended) soul and (extended) material body, the lived body ceased to be what it used to be both in ancient Greece[1] and, like a sinking stream, even in Christian anthropology prior to its surrender to Platonism. That is, it ceased to be whatever one feels in the vicinity (but not always within the boundaries) of one's material body *qua* belonging to oneself, without drawing on the five senses (in particular, sight and touch) as well as the perceptual body schema. Thus, the felt body ceased to be a "crowd" of felt-bodily isles that – impossible to exactly locate – come and go, become obtrusive, and recede into the background again. Such isles are sometimes relatively stable (oral cavity, anal zone, chest, back, belly, genitals, soles, etc.),[2] while at other times they come forward or dissolve on the basis of excitement (itch, palpitation, burst of heat, ache, etc.). In any case, they generate in the felt body a constant intracorporeal communication, a ubiquitous embodied communication with the outside world, thanks to bridging qualities that can be felt in one's felt body and also perceived in encounters with other things and persons, whether at rest or in motion. Such felt-bodily communication (cf. Schmitz, 1978, pp. 75–109; 1989, pp. 175–217; 2002, pp. 54–64; 2005a, pp. 168–184; 2011, pp. 29–53), whose effectiveness can be easily experienced in cases of collective emotional contagion (which may also be memetic), varies depending on the form of binding of the two poles of vital drive (tightness/wideness).[3] It also changes depending on whether the felt body, by means of privative expansion and privative contraction, aims towards expandedness or contractedness.[4]

Without this felt body, the outside world would only be accessible through the senses, whose synthesis has been traditionally located both in the heart and in the brain but, due to the current neuromania, is today only placed in the latter. The senses would thus be entrusted with the task of deciphering, combining and capitalizing on the signals that are supposed to come to the inner world, allegedly being able to form an acceptable and adequately manageable image of the outside world. Western culture has indeed become unable to understand the absolute and affective location of the felt body (based on irreversible directions) and the potentialities of its being-in-the-world thanks to its ubiquitous embodied communication (whose extreme poles are encorporation and excorporation). As a result, it has solely focused on the physical body,[5] which is placed in a locational space where everything is relative to mutual determination in terms of positions and distances. This material body can be easily quantified in an object-like way, to the point that it can be perfectly described in the third-person-perspective. Constructed on the basis of surfaces – in which there are points that can be connected by reversible lines – the material body implies, in turn, a single locational space, now radically foreign to the more original surfaceless and pre-dimensional space of the felt-body.[6]

b) The second repression – also connected to psychosomatic dualism and, if you like, to the so-called "discovery of the mind" (Bruno Snell) following the oblivion of the felt body – concerns feelings and their specific spatiality. In fact, psychologism consists precisely in translating/reducing the entirety of personal experience to the psyche (or soul, *mens*, mind, conscience, and today the brain): that is, to a closed-off private inner sphere. Instead, New Phenomenology frees emotions and feelings from their introjection into the soul and considers them as felt-bodily moving forces that are spatially poured out in a lived, pre-dimensional, pre-geometrical and surfaceless space (like that of sound, of weather, of light, etc.). This is how Schmitz – followed also by Gernot Böhme[7] and others – philosophically inaugurated a thematization of atmospheric feelings that today raises growing scientific (and interdisciplinary) attention.[8]

Be they pure moods (contentment and despair *in primis*) or centred excitements, what matters for New Phenomenology is to reiterate that feelings and emotions are not mainly private states of the soul, but atmospheres. In other words, they are non-physical forces – in many ways analogous to the primal images (*Urbilder*) conceived by Ludwig Klages as daemonic powers – that cannot be explained as intentional acts. In fact, Schmitz reinterprets this seeming intentionality in the terms of Gestalt psychology (Metzger), distinguishing within feelings a sphere of condensation (where their characteristic features are gathered) and an anchoring point (from which they issue forth as Gestalt). In a nutshell, atmospheres are centripetal and external vectors; they have authority and, generating affective involvement (an embodied experience), they grip one's vital drive, suddenly taking over the perceiver and leaving him based on their specific resonance[9] in his personal embodied disposition.

For the ancients, rage, eros and parrhesia – for instance – were not private inner emotions but atmospheres, partially personified respectively by Ares, Aphrodite[10] and the Holy Ghost. Also, they were taken to be so external as to be impervious to the human projective impulse, while today such impulse is regarded as constructivistically omnipotent, leading to the erroneous belief that the world is semantically and affectively empty.[11] Indeed, be they swift or gradual,[12] atmospheres govern our thymic life. In fact, it is undoubtedly true that we can simply perceive an atmosphere without being in its grip, and it can even happen that an emotion (*qua* atmosphere) is in the air without anyone feeling it; however, atmospheres normally grip a person as half-entities (quasi-things),[13] thus arousing a natural and very precise suggestion of motion, luckily not yet fully repressed by our personal emancipation.

For Schmitz, as we have seen, the capital sin is introjection:

> for every individual endowed with conscience, the world is split between that individual's external and internal worlds, with the proviso that the subject will become maximally aware of an object of their external world only insofar as such object has a proper representative within the internal world of the individual.
>
> (Schmitz, 2007, p. 14[14])

Introjection makes it inexplicable how the subject can "leave" such a prison house (box) – closed off and hierarchically articulated in layers ever since Plato – and access the outside world. As we have seen, this view (which was born and consolidated starting from the 5th century BC),[15] by which the soul provides substantial support to all life experiences, relieves humans from the dictates of atmospheric feelings, making them independent and therefore ethically-criminally liable for what could previously be attributed to transcendent forces.[16] Hence the split, even within the soul, between subject and object – think of the Platonic *topos* of thinking as the soul soliloquizing within the soul – that for Schmitz is tantamount to saying (contradictorily) that an inhabitant of a house is also the house itself. Following the introjection of external emotions (atmospheres), eros and *phobos* became no longer external powers but simple intra-psychic specializations of pleasure and pain. Thus the world started to appear neutralised and tragically devoid of all that pathically "moves" human beings; on the other hand, the soul, thus closed-off into the confines of the body, turned into a black box from which the five senses escape always only "on parole", testifying the outside world in a weak and always uncertain way.

The possibility of self- and world-empowerment comes at a dear price. The destruction of collective situations produced the disconsolate autism (social atomism) legitimating Hobbes' *homo homini lupus* and was only partially compensated by some form of sociality – conventional (Protestant world), active-mannerist (Catholic-Jesuit world) or ceremonial (*ancien régime*). As in every great narrative, everything holds in the neo-phenomenological genealogy of the West. This is why even love lost its once indispensable ideal anchoring point (the beauty, virtue and decency of the loved one), turning into a passion devoid of external legitimacy and exclusively oriented to pleasure (cf. Schmitz, 1993; 2005a, pp. 99–111). Insecurity and self-alienation were triggered by the French Revolution and by the thought of Fichte. According to the latter, the floating or wavering (*Schweben*) of I's imagination between finite and infinite or between all facts had a great "career" in the Romantics (mainly Novalis and F. Schlegel), persuaded to be able to turn "away from" anything (recessive irony) and "to" anything (productive irony). Even Hegel admits to this, convinced that "everyone will [...] find in himself the ability to abstract himself from all that he is, and in this way prove himself able of himself to set every content within himself" (Hegel, 2001, p. 30).

This ironic-nihilistic oscillation – which explains today's search for superficial experiences (in private life, the media, sport, tourism, etc.) and only offers situations that are not seriously involving – affects modern, frustrated human beings in two senses: on the one hand, they obsessively look for something to hold on to in the hope to regain some "place in the cosmos" (state, society, Christianity, logical order, common sense, Being, etc.); on the other, they would arbitrarily enact any role to impress others without being impressed themselves (think of dandies, cf. Schmitz, 2010, pp. 111–126). For Schmitz, this ironic dandyism has a philosophical face, which can even be found in Wittgenstein's *Tractatus*: in fact, assuming that there are only objective facts, he claims that what we cannot speak about we must pass over in silence (but aren't these the most important life experiences?). However, this dandyism also has a popular face: that of letting oneself go to the increasing amount of seductive technological devices and arrogantly aiming at self-realisation: "as if a self that can be realized were already undoubtedly present" (ibid., p. 126).

c) The third repression, perfectly consistent with the oblivion of the pathic and external sphere in its reductionism, concerns the outside world. The latter has been long understood based on the generalization of a rare circumstance: the calculation of solid bodies at the centre of the visual field (Schmitz, 1990, p. 21). Mostly, the outside world has been physically "reduced" to a few selected standard types, which can be exactly measured and are functional to the predictive power of hypotheses, as well as to the fact that humans can (or believes they can) take power into their own hands (Schmitz, 2005a, p. 69). After Democritus,[17] external impressions (the so-called secondary and tertiary qualities, the heart of uncontrollable life experiences, atmospheres, the sources of suggestions of movement, synaesthetic qualities and embodied communications),[18] i.e. quasi-things, were either exiled into (aestheticizing) irrelevance or, for pragmatic reasons, converted by force and with grotesque results into discrete full-things and causal processes, such as when the wind (a quasi-thing) is reinterpreted as air in motion.[19] Quasi-things, at least partly akin to the daemonic and imaginary forces thematized by Ludwig Klages,[20] are instead ontologically different from things proper, mainly because they exist provisionally and are exhausted in their action, thus implying no previous and external cause.[21]

d) The final repression, which can be also attributed to the phenomenological mainstream, concerns situations.[22] The perceptual *prius* of our everyday life is an internally diffuse meaningfulness in the sense that, within it, not everything (possibly nothing) is singular – where singular is whatever increases a number by 1 within a given *genus*. More precisely, situations as haloes of meaning are made up – also thanks to retentions and protensions – of states of affairs (something is the case), programmes (something ought to/should be the case) and/or problems (whether something is or not). Their chaotic non-numerical manifoldness – that is, their undecidedness about identity and difference[23] – is only overcome when situations (eventually) become the object of a prosaic explication, which is as such coarser[24] than a poetic one (broadly understood) and, for this very reason, suitable for singularisation. However, what matters for New Phenomenology is that, even if the analysis resolves the initial vagueness[25] and singles out the meanings that were previously a subtle part of the internally diffuse meaningfulness of situations and concepts, it must never be more fine-grained than necessary – otherwise what follows is the pathological and unscientific pedantry of wanting to know exactly what one has competence in without analytical knowledge (*je-ne-sais-quoi*, cf. Schmitz, 2014, pp. 77–78).

Situations can be of different kinds: impressive or segmented (depending on whether the whole meaning or only part of it shows), current or state-like (depending on whether the changes can be observed in the short or long run), private or collective, etc. They appear as rich and poly-significant impressions, also in the sense that they make an impression on someone and "purport more than one might say individually" (Schmitz, 1994a, p. 13), especially of course on those who have the (for Schmitz, very "womanly") ability to understand similarities and ambivalences non-deductively (cf. Schmitz, 1980, p. 379; 1993, pp. 85–86). By introducing the concept of situation, Schmitz feels he has a tool to reconsider some now unproductive notions: in fact, the "soul" becomes the phenomenologically more accessible personal situation; the "moral conscience" turns out to be only the residual form of the external pathic sphere a situation consists of; finally, the Platonic theory of ideas (which, as such, are not criteriological)[26] should be regarded not so much as a positivistic compensation for metaphysics – as Heidegger thinks – but rather as the reductionist version of atmospheric feelings and significant situations.

Notes

1 But not only: for a look at other cultures, see Rappe (1995).
2 Several painters have represented the physical body, but there are probably still no painters representing the felt body, which cannot be seen but can be felt: cf. Schmitz (1965, p. 27 ff.) and Soentgen (1998, pp. 19–20).
3 The varying intertwinement of the two extreme poles produces all other polarities, such as contraction/expansion, tension/swelling, up to the extremes of privative contraction due to a fright (and therefore impotence), and of privative expansion due to falling asleep or dozing off (for instance after having an orgasm). Cf. at least Schmitz (1965, pp. 73–172; 1966, pp. 7–36; 1993, pp. 115–149; 2005a, pp. 150–155; 2011, p. 15 ff.).
4 Unlike the Freudian impulse, this felt-bodily dynamic is non-objectual and non-finalistic, so much so that the following unity is allegedly never static or perfect, but "cruel" (Schmitz, 1965, p. 326) because of its constantly conflicting nature.
5 While opposing this reductionist turn, Schmitz at times seems to admit that, in addition to two ontologically different dimensions such as the felt body (the very topic of his phenomenology) and the material body (the object of the natural sciences), there is also an intermediate one (lived-material body), made up precisely of the isles whose localization is both absolute and relative, acting as the intersection of the two areas.
6 Cf. Schmitz (especially 1967; 1969; 2005a, pp. 185–217), and, for the relation between space levels and types of atmospheric feelings, Griffero (2014b).
7 For an introduction to Böhme's theory of atmospheres, see Böhme (2017a, 2017b).
8 The theme recurs in many of Schmitz's writings. The main ones are Schmitz (1969, p. 91 ff.; 1977, pp. 74–134; 2002, pp. 65–76; 2003, pp. 243–261; for a recent summary 2014). It is also interesting to look at Fuchs take on this (2000, passim) and Schmitz's replies (2003, pp. 175–205; 2005a, pp. 270–276).
9 The relative subjectivity of atmospheric perception (Schmitz, 1990, p. 310) can therefore be explained starting from the different felt body disposition (bathmothymic, cyclothymic, schizothymic) and the different relation between personal idiosyncratic world and personal alien world (so that a person can be, say, extrovert, introvert and ultravert).
10 Later on, eros was privatized as individual pleasure to the point that it was believed it could be boycotted (as in Aristophanes' *Lysistrata*)!
11 Some have objected that atmospheres thus risk being reified (Hauskeller, 1995, pp. 21–31; Soentgen, 1998, pp. 59, 106–110) and that the subject risks being discharged as a mere "passenger of atmospheres" (Soentgen, 1998, p. 117). I have tried to reply to such objections both on a phenomenological and on an ontological level (Griffero, 2017a, pp. 19–53; 2014a, pp. 119–129) also following Schmitz (2014) and Böhme (2001).
12 "The way in which the feelings take possession of the lived body can be both by tacitly creeping in and through a sudden and violent infestation [...] But this in no way changes the mimic certainty described. The doubt about what is the feeling that really grips us is not at all uncertainty about the choice of gestures" (Schmitz, 2002, 73–74). Soentgen (1998, 114 ff.) rightly notes that misjudging thusly the processuality of atmospheric feelings may entail a misleading reification of them.
13 Quasi-things (for example: wind, the human voice and gaze, the sense of gravity, pain, melodies, rhythms, night, lived time, heat and cold, parts of the day, figures etc.) differ from full-things because of their intermittency – they appear and disappear, but we cannot sensibly ask ourselves where they have gone and how they have existed in the meantime – and because they do not have a three-polar causality (cause-influence-effect) but a bipolar one (cause/influence-effect), as there is no autonomous cause prior to their influence. I have proposed elsewhere a first

systematization of an ontology of quasi-things such as (among others) felt-bodily isles, pain, shame, gaze, and twilight, among others (cf. Griffero, 2017a).

14 This representative can be the result of an act of seeing or hearing, a sensation or a mental image: generally speaking, it can be a content of consciousness, even related to conjunctions such as "and" or "or", according to Husserl.

15 Of course it is not easy to establish when exactly it became legitimate to translate *psyché* (originally akin to the Hebrew *nefesh*, namely the vital breath of which humans are also made of) as "soul". For Heraclitus (45 DK), for example, "you will not find the boundaries of soul by travelling in any direction, so deep is the measure of it".

16 In this sense, Schmitz points out that the psychologistic interpretation of the motto "know thyself" is mistaken, whereas we should go back to the expression's non-introjective meaning (Schmitz, 2003, p. 363).

17 For him, *noein* no longer signifies an intelligence perceiving significant impressions (as for Anaximenes, Parmenides and Empedocles), but a supra-perceptual thought able to grasp essences.

18 It is worth recalling that Schmitz (1966) explains the evolution of styles in architecture and figurative arts through the becoming of these felt-bodily elements.

19 Cf. Griffero (2017a, pp. 9–18) and Hasse (2017, pp. 173–227).

20 Haensel's whole book (1946), despite strangely never quoting Klages, rests on the interpretation of feelings as daemons, that is, cosmic powers which are real (active) and never truly "erroneous" as such (i.e. regardless of our limited intellectual judgment). On Schmitz's relation with Klages, whom he considered "the closest of my kindred spirits" (Schmitz, 1989, p. 258), cf. Schmitz (ibid., pp. 255–287) and, more generally, Grossheim (1994) and Griffero (2016b).

21 For a wider overview of the "properties" of quasi-things cf. Griffero (2017a, pp. 9–14).

22 The notion of "situation", in the *System* (Schmitz, 1977), concretizes the more primal notion of "chaotic": it is worth noting that the latter concept was still thematised by Schmitz long before it became a customary topic of the natural sciences (cf. Gamm, 1994, p. 73 ff. and for a defence against his objections, see Soentgen, 1998, p. 152).

23 Schmitz (1964, p. 312).

24 "The prosaic explication […] aims to isolate in a situation states of affairs (relevant facts) and privileged programmes (ready for decision), while throwing away all the rest" (Schmitz, 1994b, p. 237).

25 The exaggeration of the vagueness of feelings, in any case, appears to Schmitz as a reflection of the mistaken introjectionist conception of affective life (Schmitz, 1999b, p. 288).

26 In Plato there is "beauty, which is nothing but beautiful [and] goodness, which is nothing but good" (Schmitz, 1999b, p. 392).

4 Neo-phenomenological "therapies"

So far, albeit briefly, we have seen Schmitz's *modus operandi* as a historian of ideas, that is, when he combines his phenomenological analysis with a wide-scope history of orientation (as Odo Marquard, 1991, p. 91 ff. would say). What he offers us is a critical genealogy that is not at all circumscribed to conceptual cores covering at most one or two centuries. Rather, it implicitly fuses several different approaches: that of rational reconstructions, which interrogate the past starting from our problems; that of historical reconstructions, which, showing the historical character of our problems, declare they were unknown to our predecessors; and finally that of the history of the spirit, which "is needed to justify our belief that we are better off than those ancestors by virtue of having become aware of those problems" (Rorty, 1984, p. 68). New Phenomenology embodies a degree of progressivism, implicit in its presentation as a rational reconstruction and revision of canons – otherwise it could hardly claim to identify past errors and propose different ways of thinking, perhaps derived from non-professional philosophers. However, such progressivism never completely cancels its contextualistic empathy: indeed, the reader is asked to sometimes regain familiarity with a vision prior to the time marked by the given error. New Phenomenology claims to provide a – if not the only – philologically correct description of this vision, which strangely resembles New Phenomenology itself in its essential traits!

In my view, the relationship between New Phenomenology and pathic aesthetics is mostly confirmed by the task they both assign to philosophers: to shed light – reflectively, but as contiguously as possible to life – on an affective involvement that, unlike shadows projected on the wall (Plato), should not be sacrificed to the benefit of sunlight (abstract ideas). If anything, such an involvement is to be understood, and that's why Schmitz – who devoted himself to philosophy "also" because of the emotional shock of Nazism (cf. Schmitz, 1999a, pp. 9–10) – finds a common thread[1] between the three above-mentioned mistakes (autism, dynamicism,[2] ironism), one that even led to Hitlerism.[3] Or, better, it led to "disposing, in a technical and mass form, of death for death's sake through the means of modern technology" (ibid., p. 53), through which a rationalistic and hypertrophic dynamicism (mistakenly) believed to prevail over autistic deviation and even over death (the final residual obstacle to the Western will to power). This claim of dominion over death should be also seen behind the Holocaust. The latter – with a politically problematic choice, no doubt – should therefore be traced back from the Jewish genocide to its moment of incubation, namely the unaware historical experiment of the First World War;[4] but it should also be projected forward in its epochal role as foundation for the atomic rearmament.

So, in a way, Schmitz takes Hitler as a figure of the spirit, who found the cure for the autistic error – of which he personally suffered by constitution – in the ideal of the National Socialist people community. And Hitler allegedly did so in the service of the revenge (incarnated by him) in which imperialism, responsible for the First World War, found its outlet, opposing the barbarization of the dynamistic error with the barbarization (in its triumphal form) of the phobic-Faustian anti-triumphalism (Schmitz, 1999a, p. 391). Such a choice can certainly arouse some historical-political perplexities. However, it cannot be denied that Schmitz offers a great narrative that, like the many others that preceded it (e.g. those by Hegel, Spengler, Klages, Heidegger, Lukács and Foucault, to name just a few), is worthy of discussion not in spite of, but precisely because of its disproportionate ambition – at least for those who in philosophy do not only seek sophisticated analyses of minute logical-linguistic questions, but also "scenarios", no matter how philologically rash, within which to somehow place themselves and their problems.

As you can see, in this critical genealogy and history of orientation, Schmitz is not contented with studying temporarily circumscribed concepts, but – hermeneutically "examining" the past starting from "our" problems – he suggests a wide-ranging overhaul of the traditional way of thinking. Mostly, he doesn't merely diagnose historical-spiritual mistakes:[5] he also proposes a therapy to fix them. The dynamistic error, produced by introjectivism, is treatable through the (re)discovery of primitive present/presence as the centre of a life devoid of utopian alibis. The autistic error, especially dangerous in practical-existential terms, is treatable theoretically through the repudiation of nominalism and constellationism; it requires deep implanted situations[6] to be regenerated in a superpersonal *nomos* based on the unity of duties and rights, and in atmospheric feelings that free their bearers from the disorienting freedom of indifference. Finally, the ironist error is fixable thanks to the discovery of radically subjective facts.

This last point, predictably very relevant for our pathic aesthetic, deserves to be further explored here. As paradoxical as it may seem, in New Phenomenology a radical anti-psychologism coexists with the re-evaluation of subjectivity, more precisely of the rigorous subjectivity[7] discovered by Fichte.[8] The latter does not amount to "a position on the terrain of objective facts (positional subjectivity)" – earned (or extorted) by abstracting from objective facts a transcendentally privileged position, which as such can also be referred to by others (as is the case with the Kantian Copernican revolution) with equal knowledge and linguistic competence – but to "a factuality of another kind, the factuality of subjective facts for someone" (Schmitz, 2005b, p. 6).

Schmitz here is thinking about the facts that only the subject involved can express, which convey a surplus of meaning (affective and felt-bodily) that is completely absent in neutral facts, or in subjective facts derived by abstraction. This is why it can never be adequately described through the analytical distinction between one's own perspective and that of others, both in principle confined on the level of objective facts alone.

This notion of subjective facts, which only sounds like a *contradictio in adjecto* within a mindset colonized by the natural sciences and oblivious to our constant affective involvement, brings out the proto-identitarian and non-singular-numerical dimension – a real "mythogenous idea" (Stepath, 2006, pp. 119, 123) – that Schmitz defines primitive presence/present already in the first book of his *System* (1964). This presence is certainly destined to emancipate itself, articulating the five initial orientative dimensions (here, now, being, this, I) in the consequent binary oppositions[9] and thus allowing the subject to reach the propositionally accessible singularity. However, it must remain the permanent and indispensable substrate of conscious life: what one occasionally regresses to in immediate involvement (or in catastrophic situations like laughter and crying), and thanks to which, only by making possible a self-awareness that cannot be explained in the inevitably bad infinite (*schlechtunendlich*) form of self-attribution, we acquire our own profile and reliable consistency. It is therefore evident that subjectivity does not serve here as a *refugium* for all that escapes the reductionist razor, but designates subjective situations (Schmitz, 1994a, p. 15) with extreme pre-abstractive precision: that is, the situations which the subject is not added to *ex post*,[10] but pathically participates in.

Without embracing self-presentations that are a bit too emphatic to be hastily accepted,[11] it seems useful to quote in full, also and above all for the purpose of my project of a pathic aesthetics, a more sober programmatic declaration by Schmitz. According to him, New Phenomenology should

> allow self-contemplation to cast a – conceptually accurate – glance at the room for manoeuvre due to personal emancipation and personal regression between primitive present/presence and the unfolded present/presence; at the sentient lived body as a medium for every resonance and as a creative force endowed with specific spatiality and dynamics; at embodied communication as the source of any contact in encorporation and excorporation; at felt-bodily and surfaceless spaces underlying the network formed by relative places; at feelings as gripping atmospheres; at situations in their different types and several stratifications, including especially rich impressions (impressive situations). [It has to] give back to human beings, thanks to understanding, their spontaneous life experience.
>
> (Schmitz, 2005a, p. 57)

But one thing must be clear: the solution to the error of the psychologistic-reductionist-introjective paradigm is certainly not to pretend, proto-phenomenologically, to ignore the influence of tradition (Schmitz, 2003, pp. 6–7), or to regress to archaic thought, sacrificing our "analyzing, argumenting and criticizing lucid rationality" (Schmitz, 1999a, p. 403). Rather, the solution lies in a healthy critical revision – a heterodox philosophical "surfing" if you will (Schmitz, 2005b, p. 107) – that while bringing out something repressed, implicitly attributing to it a compensational and emancipatory function with respect to the prevailing theoretical diktat (Schmitz, 1969, pp. 406–407), ultimately mocks "the strange illusion that one always ought to do something" (Schmitz, 1965, p. 601). Indeed, next to and above activistic pride, the emphasis should also fall on the passive pride with which one is exposed to atmospheric feelings, to presences to get anchored to, without necessarily referring to the prosthesis of the future – let alone if the latter is conceived harmonistically (Schmitz, 1999a, pp. 380, 394).

Differently put, our will to live should learn

> to anchor itself to the present through a tenuous balance, instead of evaporating into projections and projects become vacuous in the meantime. It should certainly strive for a better future and better living conditions, but not so that the agreement with the present life should need, like a crutch, a hope in something that is yet to come. Instead of this something, one should live the present/presence (primitive and unfolded) with greater intensity, in the manner of those who are in love or dance, fully experience their dwelling or courageously exercise justice. Phenomenology cannot bring about this life form but, ensuring understanding of experience, it can prop up against collapses (as if it were the gallery of a mine) the depth of the present/presence in terms of a vision able to give an account of itself.
>
> (Schmitz, 2003, p. 19)

Because of the polemic aspect and the nonconformist anti-intuitiveness of many of its theses, New Phenomenology is somewhat revolutionary and in any case hardly leaves one indifferent. Exposing phenomena that would show themselves only if we were to re-learn to look beyond today's dominant rationalistic and technoscientific *habitus*, this approach brings together, obviously with alternate outcomes and controversial solutions, the rigorously descriptive purposes of phenomenology with the demands of a philosophy understood as criticism of culture. And the mission of the latter lies here in giving the world back some of its archaic "colour" – "how flat and boring the world would be without quasi-things!" (ibid., p. 15) – challenging the reductionist and Enlightenment dogmatism that unknowingly dwells in all of us.

As much as one may question the choice to present once again the Western tradition as monolithic and unitary, rather than considering it a porous flow sometimes inhabited by an irreducible plurality, the desirable multivocity of the narrative sciences is realized, I believe, not so much in the forcibly ecumenical mixture supported by the philological scrupulousness of perspectives within a single narrative, but in the open-minded dialogue of different, and healthfully unilateral, frescoes of history of ideas. Lacking alternatives, we trust the sciences of the spirit but do not believe them.[12] To exaggerate a little, this may also apply to the great narratives of the history of the spirit such as the neo-phenomenological one. As we have already seen several times, and especially in the last two quotations, the approach of New Phenomenology and of pathic aesthetics seems particularly fruitful in the analysis of (too underestimated) involuntary (and therefore unexpected) life experiences and their atmospheric scope. But before tackling some examples of the atmospheric approach, it is necessary to focus on some central phenomenological questions already touched on in previous pages such as the (in?)authenticity of feelings (atmospheric and not), absolute subjectivity, the externality of expressive qualities and the (rediscovered) centrality of "presence".

Notes

1 The warning signs of whose mindset are even found in the following maxim by La Rochefoucauld: "La magnanimité [which Schmitz translates as megalopsichia!, 1999a, p. 43] est un noble effort de l'orgueil par lequel il rend l'homme maître de lui-même pour le rendre maître de toutes choses".

2 This excludes – hence a less systematic push to self- and world-empowerment – both the Eastern Church, influenced by Neoplatonism and by the writings of John, and Islam, for which divine power remains very unpredictable (Schmitz, 1999a, p. 42).

3 Going through Baconian triumphalism, Lutheran individual salvation, the combination of technology and imperialism, intramundane asceticism, the oblivion of practical and ethical habitualities, and last but not least autism, poorly compensated for in totalitarianism (and in Hitler's specific phobic-Faustian anti-triumphalism) by the identification in a collective destiny favoured by a subtle technique of impressions.

4 For a critique of this genealogy of Hitlerism, understood as a hyper-compensation for the spiritual deficits of the West, cf. Landkammer (2000) and Heubel (2003).

5 For a hyper-synthetic version, cf. Schmitz (1999a, pp. 390–396).

6 Schmitz mentions Confucian ethics, foreign to the Hegelian dualism of ethics and morality, but rigorously condemns any artificially homogeneous situation (from Jacobin stagings to Nazi parades). This explains why, despite his conception of art in terms of "distanced involvement" and controlled cultivation of atmospheric feelings (Schmitz 1990, pp. 455–497; 2007, pp. 81–93; 2012c, pp. 188–192), he believes, *contra* Böhme's aesthetics (we will get back to this), that what can be intentionally produced are not atmospheres but only impressions of propaganda (Schmitz, 2003, p. 256). This is a land where those of us who profess a pathic aesthetics (cf. Griffero, 2016a) obviously cannot fully follow him.

7 As opposed to other attempts to explain the psychic apparatus, for example by framing it into rigid faculties (Scholasticism, Kant) and exemplifying it on solid bodies (Freud) or on the vortex-mechanical model of atoms (associationism, Herbart).

8 "I write something, and hence I have a representation of my act of writing. But the person next to me is also writing something. How do I know that my act of writing is not that of another person?" (Fichte, 1992, p. 425).

9 Absolute felt-bodily spatiality vs. relative space; absolute moment vs. positional time; being vs. non-being (or only possible being); personal subject vs. the other, etc. For Schmitz these dualisms, among other things, are the condition of possibility of the (always latent) hysteria in the human being (cf. Stepath, 2006, p. 124).

10 Indeed, "the word 'me' is to be understood not so much as a pronoun but rather an adverb (like 'here' and 'now'), which does not nominate an object but characterises a milieu – just as with the word 'here' we do not refer to an object ('the here') but to what is here, in the milieu of maximum proximity" (Schmitz, 1994a, p. 15).

11 New Phenomenology would be "the only attempt in the history of philosophy which succeeded in overcoming the psychologistic-reductionist-introjectivistic objectification (the source of the dynamic and autistic error) not only as a claim but as a fact" (Schmitz, 1999a, p. 403).

12 According to a very smart joke by Marquard (1991, p. 91 ff.).

Part Two

Atmospheres, presences, expressive qualities

5 Are there inauthentic atmospheric feelings?

The overestimation of intentionality

It could be said that feelings have an organic-biological basis and develop in the psychic interior of the subject, in a form that can be more (fear) or less (envy) culturally invariant, thanks to a certain interaction with the environment. But this thesis is too easy to be true. First of all, sometimes we do not "feel" well and yet we are unable to name – and therefore pre-interpret, classifying it typologically – our affective state; its expressive resonance only lies in the experience of it, which is felt-bodily (*leiblich*) but not necessarily also anatomic (*körperlich*). Also, in our culture feelings are considered something private,[1] even when they seem to perfectly correspond to an economy based not on natural needs but on necessarily unfulfillable desires (Böhme, 2017c, pp. 9–31). Our daily experience, on the other hand, seems to suggest that, in many if not all cases, the affective is rather a spatial quality of the external world: as we have already mentioned several times, it is an atmosphere.

Of course, in this radical enhancement of the emotional, it is better to take some precautions. Abandoning Descartes, provocatively claiming that "we feel and therefore we are", does not mean in fact confusing the rediscovery of the pathic dimension (see Chapter 1) with today's vulgar media psychologism, fundamentally obscene in its spectacularization of affective states; nor does it mean mistaking it for the late-capitalistic ubiquitous tendency to emotionally design any consumer strategy.[2] Rather, it means taking seriously the *affective turn* that occurred in many human sciences,[3] trying to establish its foundations based on a neo-phenomenological atmospherology. Feelings are "powers", which are atmospherically effused into external space,[4] responsible for a felt-bodily communication able to resonate in the perceiver's lived body[5] and to authoritatively take over.[6] Admitting this also means acknowledging the need to reevaluate the archaic conception of the affective as a daemonic-extrapersonal and spatially diffused antagonist,[7] *contra* the now prevailing psychologization and the projectivism that follows from it. And this, of course, does not mean dreaming of an impossible regression to a pre-introjective life form, perhaps driven by the obscurantist intention of condoning emotional life, but aiming, more simply, to a healthy rebalancing of the prevailing psychic ontology.

Thus understood, atmospherology cannot be content with recognizing the superindividual (social, pragmatic, intentional, normative) implications of the feelings experienced by the subject. Interpreting their "felt-bodily synaesthetic resonance" (Fuchs, 2000, p. 214) neither metaphorically nor causally – walking chin up is not "caused" by pride, it *is* pride! – it cannot be content with a cognitive re-evaluation of feelings either.[8] In fact, if it did so it would end up misunderstanding the role of the felt body, i.e. the analogy between the expressiveness of the states of the world and that of "felt-bodily existentials" (or apriori of the lived body), which can be defined "somatization" only very reductively. Also, tracing emotions back to dispositions (Wollheim, 2001), it would end up fatally underestimating their sudden and powerful insurgence. On the other hand, atmospherology takes everyday (especially involuntary) life experience seriously and wonders "what it is like to feel what one feels". In so doing, it rejects both the genetic question (in what series of events does this feeling occur?) and the causalist one (what is the cause of this feeling? The "usual" amygdala? Cf. LeDoux, 1996). It rejects both the functionalistic-adaptive reductionism (for which, for example, one would feel ashamed *so as to* avoid being attacked!) and multifactorial theories, fundamentally incapable of establishing how and by virtue of what hierarchies the various factors give life to the holistic emotional totality that we experience (Blume and Demmerling, 2007, pp. 127–128). Despite their sophisticated demarcations,[9] such theories do not sufficiently emphasize the fruitful semantic vagueness of the notion of emotion[10] and of the following family resemblances.

In short: atmospherology is undoubtedly part of the philosophical rebirth of affectivity, but its approach – like that of so-called folk psychology – must be limited to the dimension of *Lebenswelt*, epistemically impenetrable in its perceptual nature (cf. Döring, 2011), and *a fortiori* irreducible to empirical-experimental research.[11] Physicalist "reductionism" turns out to be a real categorical error, since feeling something does not mean taking an interest in neuronal processes: the latter may even be the cause of the feeling, but certainly they are not phenomenologically perceptible (you can say anything about the brain, except that it has emotions in itself!). Therefore, atmospherology is much more radical than the rediscovery of the "also" cognitive and rational role of feelings,[12] even without embracing either the behaviourist equivalence between feelings and behavioural forms, or the suggestive one between feelings and physiological processes (James-Lange theory).[13]

Except for "groundless" atmospheres (Schmitz) – that is, atmospheres devoid of a likely point of origin – as quasi-things or effects of the "ecstasy" of things, atmospheric feelings are relatively homogeneous complexes of affordances (Gibson, 1986). They are irreducible to occasional subjective vibrations, but usually can be attributed to "generators" (Böhme, 2001, p. 87 ff.), whose effectiveness may even be, within certain limits, counterfactually planned. Ephemeral and intermittent, yet endowed with an identity that makes them, within certain limits, recognizable and linguistically expressable, atmospheres are pervasive ways-of-being (Metzger). They tonalize the affective space in which we (literally) enter and which we deal with, segmenting it through non-geometrical boundaries. They can capture us resisting any attempted projective transformation, they can be noted but not shared, they can go unnoticed or even be changed by someone with an opposite and more intense mood: whatever the case, in most (i.e. non-prototypic) cases, they are an "in-between" between a subject and their surroundings (things or people) – an "in-between" that may even be prior to the rigid constitution of these two poles.

But above all, as these experiences are not *exclusively* introspective, atmospherology does not need to establish their specific intentional object or "formal object" (Kenny, 1963, p. 189) to diversify them as instances of some emotional "ideal type". To feel the atmosphere of envy around someone does not only mean to "judge them" worthy of envy, to "evaluate" them as having the non-descriptive characteristic of enviability. It also, and first of all, means to felt-bodily experience a (centrifugal) direction that impregnates the space with something that is still confusedly "enviable". This atmospheric space is therefore irreducible to a) a formal (intentional) object, as we shall see better shortly; b) to a propositional evaluation, as there are also anaffective evaluations or non-judgmental cognitive performances; c) to a merely dispositional-potential property, as such inferable only on the basis of some "active" emotion; finally d) to a mere (sociological) function of "role" or of informational type (cf. Nussbaum, 2001). Being interested in the *qualia* of affective experience, which are very different from pure mental states (disembodied) and from pure (unnoticed) physiological processes, atmospherology does more than simply valorize the so-called feeling-component within an intentional notion of emotion. So – leaving aside a large number of problems[14] and legitimate objections[15] – I will now focus on two points: the (supposed) intentionality of (atmospheric) feelings and their possible inauthenticity.

Precisely the spatial character (again, in a pre-geometrical sense) of atmospheric feelings forces me to start by clarifying the thesis of their unintentionality. Affective experiences of consciousness are not intentional (at least not initially), and in any case they are not intentional in the punctual-atomistic sense of cognitive acts but in that of a reference to an enveloping totality. This thesis was possibly already expressed by Husserl[16] – it is up to the philologist to decide. And yet this legitimate perplexity has not prevented phenomenological orthodoxy from uncritically agreeing with the cognitivist approach in proclaiming the intentionality[17] of "all" states of consciousness, including feelings, thereby resorting more to the mind than to the body (let alone the lived one). The result is an unsatisfactory oscillation between the prepersonal plane (physiological or neuronal) and the purely personal one (mental state, intentionality, etc.), or an equally unsatisfactory progression from a physical process to a mental one (or vice versa), within an experience felt as unitary (Demmerling and Landweer, 2007, p. 21).

Take, for example, love and respect. Rather than a state of consciousness directed to an external object (the loved or respected person), this feeling is experienced as a situation imbued with an enveloping atmosphere, in which the beloved person is certainly inscribed, but which is actually prior to their singularization, only possible thanks to the typological identification of the vague initial emotion required by intentionality. The direction of the latter should indeed be reversed: challenging the lazy reason induced by the grammatical prevalence of active verbs, we should in fact imagine here an arrow that goes from the object to the subject, who in a certain sense is "extorted" a certain felt-bodily resonance (in our example: desire and esteem).[18] Sometimes, according to the Gestaltic foreground/background logic, from the chaotic multiplicity of affective stimuli something more focused and more circumscribed emerges – therefore confused with the intended object. This could be explained, according to Schmitz, through the "centring" processes.

As the temporary and fragile focus of a multifaceted and diffused affective field, which determines it and by which it is in turn determined, the centring of an atmospheric field seems to explain our daily affectivity better than a rigid subject / object intentional dualism. A widespread atmospheric feeling, in fact, can contract until it is condensed into one or more objects, erroneously interpreted as the generators of that feeling. This is why a lover loves (atmospherically) even the doormat of their beloved's house.[19] Sometimes the intentional object is therefore only apparent, because, as Schmitz noted, the centring may not be found in the anchoring point of the percept (its generative point) but in its sphere of condensation (where the character of the percept is concentrated), so that it is not unusual to find the anchoring point "at the *opposite* end *of the condensation zone*" (Metzger, 1941, p. 230). The aforementioned doormat of the beloved's house, surrounded by a positive atmosphere, is obviously only part of the sphere of condensation of love, yet is sometimes more intensely felt than the concrete presence of the beloved person (who is the anchoring point). Just as one usually does not look at the sun directly but only at the things enlightened by it, one often confuses the (perhaps occasional) condensation point of an atmospherically diffused feeling with its intentional object (Schmitz, 1969, p. 322).

Certainly this is not the right place for a critical discussion of Schmitz's distinction – a real "hand-to-hand combat" with the affective sphere – between pure feelings (fulfilment / despair, nostalgia, or the spring mood), which, unlike commotions (always directed from tightness to wideness), are moods lacking a precise orientation,[20] and other directional feelings-moods, whose extension is, however, based on the kind of extension which is typical of pure feelings[21] (with their fluid cross-border areas). However, I do claim that the atmospheric feeling – for example the collective and spatially widespread worry in the age of global terrorism, which condensates even in inappropriate (if not random) things, people and events – is neither a pure feeling devoid of condensation[22] nor a fully intentional feeling. In fact, it is difficult to explain an atmospheric feeling in intentional terms, both if we consider it when it catches the percipient,[23] and when we perceive its complex structure, temporally articulated and also culturally channeled.[24]

This explanation of apparent intentionality in terms of centring/condensing may not be exhaustive,[25] but certainly helps outline the very common experience of vaguely and holistically feeling atmospheric affordances without managing to focus on a single element. Without going as far as Mozart's insatiable Don Juan, is it not true that one does not only love a person as such but rather (or also) the situation that they temporarily, accidentally and often even inexplicably condense?[26] The atmospheric nimbus effused in the condensation sphere is by no means "purely presentational", as Husserl erroneously claims (2001, II, p. 111) when speaking of an event "clothed and coloured with sadness". Nor can this nimbus be deprived of its initial chaoticity in any circumstance by means of progressive centring (either thingly or personal) intended as its objective end. In short: we can very well feel threatened without identifying a single threatening element. When we feel an atmosphere, an absolutely subjective fact is being expressed as the resonance of a situation that cannot be reduced to an intentional object.

The second question is whether or not there are false[27] or inauthentic atmospheres. To be clear: the question is not whether one can make a mistake in producing an atmosphere, generating for example an unwanted one – this happens very often due to the always looming heterogeneity of ends, and atmospherology should always take it into account. Rather, the question is whether one can be mistaken in perceiving-feeling them, especially at the "first impression" level. And here the answer is far from obvious.

a) The atmosphere I have called prototypical is neither false nor inauthentic. It grasps the percipient, completely redirecting their emotional situation and being impermeable to any corrective projection.[28] Indeed, for the authority and the discrepancy with which it imposes itself, it is the atmosphere *par excellence*. And the fact that sometimes one undertakes to mobilize all the critical resources available to resist such unexpected atmospheres demonstrates more than any other experience that one has correctly (ingressively) perceived an external and non-subjective feeling. As an involuntary first impression, in fact, the prototypical atmosphere should not be confused with that which arises from a possible and more analytical subsequent perceptive process whose parameter – note well – is still the first impression.

b) The atmospheric feeling that "from the outside" could appear as the result of a distortion is not false either. The atmospheric indignation that, for example, masks an atmospheric envy is not false only because an extra-phenomenological and third-person (e.g. psychotherapeutic) analysis[29] can explain it as a hyper-compensation for a sense of inferiority and as the repression of a socially stigmatized feeling. Likewise, the centrifugal anger that captures those who are "actually" feeling (and reacting to) an unmanageable sense of shame is not false. The fact that adhering to a real or apparent feeling depends on the oscillation between what Schmitz calls different levels of personal emancipation (see Chapter 6) does not at all imply that the distorted one is felt inauthentically (i.e. only half-sensed).

c) Atmospheric feelings are also not false when they lack a specific object or when, as in the case of spurious atmospheres, they can have any one object, as long as it satisfies the projective desire of the percipient. A paranoid person, in fact, does not consider their mood inauthentic, nor would they be content to explain this arbitrariness with the dialectic, already mentioned, between anchoring point and sphere of condensation.

d) Even an atmospheric feeling due to a perceptual deception does not seem to me to be false or inauthentic. Consider the silence that the newcomer deems tense and which later turns out to be simply a momentary pause in the discussion, or the harmony that is later understood to be only the "quiet before the storm": these feelings are neither false nor inauthentic. Regardless of the nostalgia for what is supposedly "natural" or "spontaneous", the same can be said of an atmosphere generated through the deceptive manipulation of materials and effects, or of a pharmacologically induced one (for example by means of an euphoriant drug)[30]: in fact, the "abstract" production of affectivity (as shown by the Nuremberg parades and by every stage) can turn out, obviously for those who have not planned it, even pathically more effective than a "spontaneous" one.

e) Also, an atmospheric feeling is not false or inauthentic just because it only has practical and mental effects on some of those exposed to it. As shown by the Kantian subjective universality of the judgment of taste, a state of mind – atmospheric, in this case – does not need to be objectively legitimated and therefore judged authentic by a statistically ascertained inter-subjective plausibility.

f) The same holds for the feeling due to the mimetic contagion of emotions that others experience before us and that, acting as precedents, have sometimes crystallized into a tradition (novels, films, history, etc.): this feeling is not false. Indeed, our whole emotional life can hardly prescind from human interaction and the inevitable mimetic relationships that derive from it, including affections that seem natural only because they are embodied in long-lasting public practices.

g) An atmospheric feeling cannot be considered false just because it is unsuitable to generate corresponding behaviors. For example, a shabby and uncertain behavior, far from being incongruous, could be the "right" felt-bodily resonance of a specific atmosphere, in this case non-thetic and of low intensity.

h) Similarly, an atmospheric feeling that never becomes fully conscious is not false. In fact, many atmospheres are typically perceived peripherally compared to the focused percept, and yet this non-theticity does not affect their influence on the perceiver.

i) A vicarious atmospheric feeling, felt "in the place" of someone who should but doesn't feel it, is far from false. Think of the emblematic and already mentioned case of vicarious shame (Griffero, 2017a, pp. 79–92): like any other atmospheric and therefore supra-personal feeling, it communicates felt-bodily with the bystanders, involving them in spite of the indifference of those responsible for it. Another example is the self-destructive sense of guilt that is sometimes felt by the victim of an abuse and not by its perpetrator.[31]

j) Finally, an atmospheric feeling is not false just because it sounds dissonant with respect to the biography, so to speak, of the lived space and the situation it pervades. Think of unexpected joyfulness at a funeral, a party so dull as to sadden its guests, or a day so beautiful as to sharpen a depressed person's sadness: these examples show very common experiences that, highlighting an emotional contrast, are extremely useful to further demonstrate the objectivity-externality of atmospheric feelings.

And the list could go on. The point is that if someone believes they are happy (in our case: if they are feeling a happy atmosphere), they are. If not, not. Therefore we cannot claim that there are authentic and inauthentic atmospheres (which should be distinguished from *bad* and supposedly manipulative atmospheres). Indeed, when we notice an atmosphere without being affectively involved by it for various inhibiting factors (over excitement, worries, unsuitable psychophysical conditions, etc.), that atmosphere is nonetheless authentic. Also, a distinction between authentic and inauthentic atmospheres, being implicitly normative, would entail a third-person *adaequatio* parameter, as such very far from a first-person and neo-phenomenological approach and from the affective sphere in general: in the latter, if really felt, a feeling cannot ever be false or only illusory (… compared with what?).

In fact, consider the belief that an atmospheric feeling is authentic only when it is justified[32] by some state of the world grasped on the representational level and physicalistically legitimated (De Sousa, 1987, p. 164), or the view that an atmospheric feeling is only authentic when it turns out to be strategically functional in an instrumental-evolutionary sense. If this were the case, much of the emotional sphere would have to be considered pathological. Instead, it is extremely common to find situations in which you feel that-p even knowing that-non-p (is unrequited love false?), or situations that end successfully despite being guided by unjustified emotions. A normative approach to affectivity, by presumptuously degrading phenomenological authenticity to mere psychological sincerity, appeals to the far from obvious idea that everything that can be amended should be amended (Salmela, 2006, p. 396). This idea, in fact, forcibly leads one to consider cognitive impenetrability as such pathological. Yet the latter exists in ordinary affectivity: for instance, we often continue to feel a certain way even though we have no reason to (at least no external reason, because evidently we have internal ones). Also, cognitive impenetrability is found in the most common perceptual illusions, like that by which the stick immersed in water continues to appear broken despite our knowledge that it isn't. The error immunity of atmospheric perception-emotion does not seem to me to be threatened by the performative contradiction: we cannot "believe" that-x and at the same time that-non-x, but it is indeed very normal to felt-bodily "feel" that-x, while knowing consciously that-non-x (Döring, 2011, pp. 267–269).

Of course, one may sometimes have doubts about the object of one's atmospheric feeling, but never about the atmospheric feeling itself (excluding pathological cases). This means that what is authentic is not the percept but the perception. Minimizing the emphasis usually placed on authenticity,[33] a pathic and neo-phenomenological approach is committed to (non-paradoxically) weaving together the maximum externalism of atmospheres and the maximum subjectivism of their felt-bodily resonance. From this standpoint, an atmospheric feeling is authentic insofar as it is felt-bodily experienced – regardless of its causes and even its effects – in the form of a first impression that is personally involving and can only be amended *ex post*. In fact, while the extrapathic perceptive truth can be in principle verified independently of the percipient, the pathic truth is at least always co-constituted by a pathic subject impermeable to "expert" evaluations (naturalistic, sociological, statistic, etc.). The reference to "reasons" does not seem decisive to me. To make an example, the atmosphere of fear for a terrorist attack at the airport may certainly seem legitimate only when there are good reasons to consider it probable: however, these good reasons could be valid for one person but not for another. Above all, the acknowledgment that there has been no attack is not resolutive, as the atmosphere of fear does not refer so much to what happens as to what "could" happen and therefore can never be denied.[34] The increase in security procedures in the aftermath of an attack does not help either, as the atmospheric feeling suggested by the attack is obviously still floating in that space and perhaps is improperly condensed even in the most innocuous object poles. In short, if a feeling (be it or not atmospheric) were only authentic based on the truth of the fact that triggers it or derives from it, much of our affective life would be reduced to self-deception.

An atmosphere, therefore, could be said to be inauthentic only insofar as, in the context of a third-person-perspective (by principle extraneous to the descriptive and non-genetic approach of a neo-phenomenological atmospherology), it is judged to be manipulated, unjustified, unconscious, deceptive, rationally uncontrollable, artificially induced, too idiosyncratic, etc. But its perception in terms of first impression, in the framework of a first-person-perspective, seems so cognitively impenetrable[35] that it can always be considered authentic. Can we deceive ourselves about the material cause of an atmospheric feeling, about the pragmatic adequacy of our felt-bodily reaction to it, and even about its nature (is this anxiety or tension?). Certainly we cannot be deceived, however, in the self-attribution of the affective state that derives from it, knowing that we are the ones who experience this or that feeling. Atmospheric feelings are therefore endowed with a relative objectivity (because it is only quasi-thingly) whose irreducibility to mental states located in the interior is paradoxically proven by their "absolute" pathic subjectivity. Which is what I will now clarify in the following chapter.

Notes

1 It is no coincidence that, except for some free zones (psychotherapy, psychiatry, but also, unfortunately, talk shows), it does not seem polite to speak publicly and *expressis verbis* about feelings (especially in reformed Christianity).
2 Gumbrecht is right: there is no definitive answer to the question of how "to avoid dissolving into what Hegel accused of being 'the mush of the heart' […] Nor is there a sure way to guarantee immunity" from this trivial emotionalism (Gumbrecht, 2012b, p. 16).
3 Interdisciplinarily interested especially in shared emotions (cf. e.g. von Scheve and Salmela, 2014).
4 I maintain this "collocation", despite knowing how ontologically misleading the "internal"/"external" metaphor can be (Landweer, 2011, p. 241). A good definition is that according to which an atmosphere "is *not* at all *'inside'* in some interiority […] but for this reason it is *not at all outside either* […] An attunement [*Stimmung*] is in each case already there, so to speak, like an atmosphere in which we first immerse ourselves […] It does not merely seem so, it is so" (Heidegger, 1995, pp. 66–67). But this does not mean, however, that one should consider oneself entirely exempt from the controversial topological issue of the affective.
5 This should be carefully distinguished from simply concomitant bodily events. Similarly, one ought to distinguish between blushing from heat and shame.
6 Cf. Griffero (2014c). Blume/Demmerling (2007, pp. 131–132) reductively believe this authority to be only due to the social practice that is connected to it. The fact that everything that "happens" involuntarily (or even against our will), including thoughts, authoritatively takes possession of us shows the quasi-affective nature also of certain cognitive aspects rather than disproving the authority of feelings (Demmerling, 2011, p. 48).
7 "Which embraces and permeates everything, which is suprapersonal and at the same time trans-objective or better pre-objective, not letting itself be 'placed' in any circumscribed object" (Schmitz, 1969, p. 102).
8 Not even of the narrative kind (Nussbaum, 2001; Goldie, 2000, pp. 11–49, with reference to Musil).
9 Proto-emotions or primary and non-decomposable emotions with adaptive function, cognitively-culturally more evolved multifactorial secondary emotions, passions, moods, and so on.
10 Regardless of an error that is difficult to circumvent: that of inferring something universal from some determined feelings. Perhaps even Schmitz runs this risk when he gives the impression of extending the typical authority of sadness to any atmospheric feeling.
11 Even the kind that, wishing to review the ordinary language relating to the affective, aims to test its validity in the natural sciences (Griffiths, 1997).
12 Which can be summarized in the idea that feelings would allow for faster, and therefore sometimes more functional, decisions than reflective ones, in addition to making possible rational action itself, because, being implicitly value judgments, they underline what is really significant in our experience.
13 Felt-bodily processes are not so much the atmospheric feeling but rather its resonance (Schmitz, 1990, p. 304). However, they certainly are its implementation.
14 Without claiming to be exhaustive (but see Griffero, 2014a, pp. 119–149): differentiation (by greater or lesser objectivity and authority) of atmospheres; phenomenological (and possibly transformative) course of their perception; distinction

between only sensed atmospheres and involving ones; relationship with the sphere of sensibility, of synaesthesia, of common sense and of environmental affordances; physical or "transcendental" character of atmospheric perception; ontological (quasi-thingly) properties of atmospheric feelings; problematic distinction between gnosic and pathic in admitting that it is possible to sense feelings without fully feeling them, etc.

15 For a tentative answer to the main objections raised to the neo-phenomenological approach to atmospheres cf. Griffero (2017a, pp. 19–53).

16 When Husserl (2001, II, pp. 110–111) says that an event arousing joy "seems to be bathed in a rosy gleam" and a sad one "seems clothed and coloured with sadness", he undoubtedly seems to be one step away from conceiving feelings as an immersion, relatively indifferent to the subject/object distinction, into complex units (chaotic-manifold, irreducible to discrete elements) like atmospheres. I would like to see these units as musical scores, based on parallel or discordant "voices" that are only temporarily dominant (Frese, 1995).

17 Though perhaps non-propositional (Nussbaum, 2001, p. 127).

18 But the enormously misleading arrow metaphor should perhaps simply be replaced by the thesis of the coexistence of the two poles (Wiesing, 2014).

19 When one sees a sad light envelop the city abandoned by the beloved, for example, "the object of the feeling of sadness is not the city, but the fact that the person has left the city" (Geiger, 1911, p. 19), thereby "tonalizing" it in a specific manner.

20 They are therefore groundless in this technical sense and show a full or empty extension in wideness.

21 "Pure *Stimmungen* come through all feelings" and "form the original layer, the ground or background of feelings" (Schmitz, 1969, p. 263).

22 Such as unmotivated happiness and boredom. Individual (even atmospheric) feelings are nothing, in fact, but the circumscribed and pseudo-intentional articulation of a stable, extra-intentional and non-focused affective background. This background is formed by *Stimmungen* (Slaby, 2007, p. 99 ff.; Ratcliffe, 2008) acting as a "sense of the situation" or an extra-objective reference to the world, which in any case (*contra* Schmitz) is atmospheric, since it could be precisely the different atmospheric colour that distinguishes one situation from another. For Landweer (2011, p. 237), the atmosphere is "in between" the situation and the subject and turns into a real situation only when its affordances are consolidated.

23 In the sense that the first impression, cognitively impenetrable and immune to errors (as we shall see), may well be followed by a more correct centering (Griffero, 2014a, pp. 136–139).

24 Frese (1995). The risk of constraining the authority of atmospheres to historically provisional cultural codes that have not evolved for some time (for Frese this includes lyrical opera!) is almost as serious as reducing them to linguistic praxis (Demmerling, 2011, p. 54): newborns and animals, devoid of linguistic articulation, would consequently lack an affective sphere and an atmospheric perception, which of course is not the case.

25 Is it possible, for example, that the same "object" is both the anchoring point and the condensation area in the case of "true" love? See Schmitz (1980, p. 275; 1990, p. 302).

26 Something that "flares up inexplicably within the feeling's space of wideness" (Schmitz, 1969, p. 322). It also applies to situations in which an anguishing object is nothing but the crystallized occasion for a more general anguish (Bollnow, 1956, p. 38).

27 The expression "false atmospheres" usually signifies states that (somehow) would not even be atmospheres in the proper sense.

28 This would act transformatively, in fact, not so much on the feeling but rather on the situation that generated it.

29 Schmitz's (2011, pp. 38, 41) differentiation between "conscious living" and "noticing", or between "involvement in act" and "involvement in a waiting position", seems problematic to me.

30 But is the atmosphere generated by reading a story or by the planned visit to a suggestive place necessarily more authentic than a pharmacologically induced one? The positive and promising atmosphere of my workplace felt under the effect of an antidepressant or some psychotropic substance, regardless of its negative consequences (addiction, abstinence, etc.), is no more inauthentic than that felt due to my desire for the colleague of the office next door.

31 Whereas an intentionalist explanation should presumably assume that the intentional object of the victim's guilt is exactly the guilt, however unaware, of the perpetrator (Landweer, 2011, pp. 248–249).

32 Or "appropriate", since in the emotional field appropriateness often corresponds to (propositional) "truth" in the epistemic field.

33 Without denying, obviously, its (at least minimal) necessity in the construction of a coherent social identity.

34 It is no coincidence that the atmosphere of fear suggested by the medical analysis we are about to receive vanishes, obviously, when we discover that we are not sick, but also considerably diminishes, paradoxically, when we understand what disease we should treat: our mood thus becomes more pragmatic than atmospheric. Much depends, of course, also on what is at stake. This explains why, for example, the restless atmosphere suggested by a weak pain in the chest is higher than that suggested by a much more acute pain in the big toe.

35 For more on this topic applied to emotions cf. Goldie (2000, p. 76 f.).

6 Good regressions

Mineness and absolute subjectivity

When one is lying in the sun in the summer heat or raptured by an intoxicating scent, one's self – whatever that means – becomes vague and then gets lost in the vastness, or "oceanic feeling", whose depersonalizing consequences famously intrigued Freud. Something analogous to these non-pathological forms of depersonalization can be generally attributed also to the body. To make two very different examples, the body's fluid and graceful functioning both in athletic performance and in the spontaneous ability to fall asleep is only possible when one does not focus it intellectually. Setting theticity aside, personal existence must indeed show an intermittent nature as opposed to an "almost" impersonal existence which, to put it with Merleau-Ponty, "can be practically taken for granted, and which I rely on to keep me alive" (Merleau-Ponty, 2002, p. 96). Therefore, in these and many other cases – which are certainly not philosophically irrelevant only because they are related to everyday life – the impersonal is really far from enigmatic. Rather, it simply marks the anonymousness that necessarily circumscribes the (quite narrow) personal sphere, and that can be compared to the "darkness needed in the theatre to show up the performance" (ibid., p. 115). If it is the case that "everything has meaning" (ibid., p. xxi) and that "because we are in the world, we are condemned to meaning" (ibid., p. xxii), well much of that meaning – especially that which acts as a background to discrete figures – is necessarily impersonal-anonymous, and not constructed by us. After all, phenomenology had already noted this by adding, next to the act (or thetic) intentionality, the operative (acting) and pre-thetic intentionality, which as such can only be known in its *ex post* effects.

This shows that it is undoubtedly possible to develop a philosophy of the impersonal with a low degree of metaphysical melodrama. The problem is that the Western philosophical, religious and naturalistic tradition has had it all too easy with the idea of "person", which notoriously substantializes a very private inner space that, in principle, can govern and inhibit the involuntary things that happen to it. This dogmatic reductionism certainly mustn't be opposed with the complete overcoming of the personal, perhaps by claiming that today's flexible human and professional relationships make the personal relatively superfluous. Let's not get carried away: after all, and despite the erosion of the notion of "person", the reader rightly expects this book to not have been written by some other person and, to make another example, if I committed a criminal offense, it is expected that I would be personally prosecuted for it.

As we have seen, a pathic and atmospherological aesthetics rests on the necessity of mitigating a rationality derived by excluding and removing the other, as a result of paranoia. Therefore, it is probably legitimate to expect that it would indeed rehabilitate the impersonal, further campaigning against the self and the person, degrading them to narcissistic notions required by the abstraction of Western rationality and by binding and heteronomous social needs. Instead, without claiming to dismiss, with a purely theoretical *coup de théâtre*, the banal Enlightenment conception of the full (if not paranoid) autonomy of the subject, the atmospherological approach underlines the necessity of the concept of "personal situation", both for self-consciousness and for everyday life. Indeed, rejecting the reductivism implicit in a strictly cognitivist approach, blind to what involuntarily happens "to us", the neo-phenomenological-atmospherological approach focuses on "what it is like to be (and feel)" here and now in the presence of something, and can therefore easily disobey the injunction to dismiss the idea of personality, which is today believed to be "philosophically correct".

As mentioned, from the standpoint of a pathic aesthetic, the subject is not so much autonomous but rather sovereign and mature, precisely insofar as they know how to be exposed (in the right, also felt-bodily way) to the involuntary things that happen to them. They must also be able to accept their shortcomings and dependencies instead of trying to deny them by seeking shelter in illusory states such as ataraxy and apathy, autarchy and autonomy (and, theoretically, in the effective and continuous scepticism required by the so-called "school of suspicion"). There is no doubt that concepts like "subject" or "person" have been pathologically overestimated by Modernity, with well-known negative consequences (including dualism). Such notions, rather than a substance, indicate an epiphenomenon that is not even fully in control of its own thoughts, and can therefore be released from the function of socially introjected normativity it is usually attributed. But a pathic and felt-bodily subjectivity is completely different from rationalistic subjectivity, because it dissolves the "I" into the (perceptological) "me" and the (neo-phenomenological) "to me".

Merleau-Ponty (2002, p. 250) already noted this: "every perception takes place in an atmosphere of generality and is presented to us anonymously [...] So, if I wanted to render precisely the perceptual experience, I ought to say that one perceives in me, and not that I perceive". But to acknowledge the necessarily anonymous nature of perception one has to take one step further, for example by noting that humans do not at all "access" the world, and not because the latter is hidden, but rather because humans are always already taking part in it and involved by it (also and mostly on the affective and felt-bodily level). Thus, while underlining the theoretically misleading character of the very metaphor of "access",[1] it is better to examine perception neither in its genesis or cause – which as such is metaphysically ineffable – nor in its transcendental conditions, as this results in a *philosophia* that is always and only *interrupta*. Rather, perception should be studied in its necessary and much more comprehensible consequences on the perceiver. But when I wonder "what I am like" when I perceive – that is, when I am alive, given that life and perception are coextensive – my "I" is replaced by my "me". In the framework of a pathic aesthetics, the expression "subject *of* perception" – understood as a subjective genitive – gives way to the expression "subject *to* perception". I thus realize I am a product of perception, a "me" rather than an "I": I do not first exist and then, among other things, perceive, but I exist only insofar as I perceive.

In other words, the perceiving person is essentially constituted by the impersonality of perception, as they co-belong to their perceptual situation; the latter determines the perceiver by involving them in a (literally) participative relationship that can be considered relatively impersonal for its adirectional intentionality. The perceiver is forced – in a way, condemned – to perceiving, as a result of which they exist in one way and not in others (which at most they can imagine). In short, they are subject to an impersonal level that is prior to subjectivity traditionally understood and that forces them to admit the present and unquestionable existence of the perceived. The perceiver is thus "condemned" to be certain of what they perceive in the world in which they felt-bodily and affectively take part. Therefore, they have to acknowledge the public (intersubjective) character of the perceived along with the perceiver's own identity in space and time (without which it would be impossible to notice change). In short, as summarized by Wiesing (2014, p. 129), "that I have a body, that things affect me, that I am visible, that I have a spatial and temporal identity, that I am not the only one in the world – none of these are contingent facts in the world, but *a priori* necessary certainties for a perceiver".

On this phenomenological level, still relatively foreign to an affective and felt-bodily perspective, it is already possible to say that personal identity, coinciding with what we have perceived and now perceive, appears to be rather produced by the effect that the impersonal has on the "me", though declined in a totally subjective way. To use a literary quotation:

> I should say straight out: Never tell anyone you know how she or he feels unless you happen to be, just at that second, stabbing yourself with the very same knife in the very same place in the very same heart she or he is stabbing. Because if you're not, then you don't know how anybody feels. I can barely tell you how I felt when … .
>
> (Ford, 2006, p. 222)

However, according to Wiesing, what dethronizes the I is not only the "me", which can be reached by following a mainly cognitivist reflection, but the "to me". This dative, or better medial, form, referring to an ontology (and a grammar) half way "between" active and passive,[2] undoubtedly highlights a relationship with otherness that is absent in egological solipsism. The person referred to as "to me" is mainly involved: before becoming an "I", "to me" signals that the subject is not yet properly what is meant by "person", and is touched and involved by what happens. Far from being self-founded, this sort of pre-personal person is based on something that it is not (yet) and that rather happens to them or is presented involuntarily.

In the framework of this oblique ontology, recently sketched by Gernot Böhme, one can say that whoever feels, imagines or thinks about something always starts from some involving "commotion", if only the unexpected discovery that something is not the way it seems. The mentioned medial ontology expresses the fact that, while being passive in giving in to what happens, the perceiver is also active, at least insofar as they accept to be exposed to it without paranoidly trying to avoid it. As happens in the perception of colours, "suddenly the sensible takes possession of my ear or my gaze, and I surrender a part of my body, even my whole body, to this particular manner of vibrating and filling space known as blue or red" (Merleau-Ponty, 2002, p. 246). When perceiving colour, to stick to this example, I am depersonalized, going back to the anonymous and operative impersonality that preceded my I. However, upon closer inspection, this is paradoxically the impersonal conditioning of an absolute subjectivity. In fact,

> perception only becomes subjective when the pathic element comes into play, that is, insofar as something happens to the perceiver in perception, if they perceive something and are perceptively involved. In fact, it is in involvement and only in it that I am irreplaceable. Others can replace me in my actions, but not in what happens to me and I perceive. It is also for this reason that, as the root of subjectivity, the "to me" is more fundamental than the I
>
> (Böhme, 2001, p. 123)

If perceiving implies the perceiver's self- givenness, that is to say being grasped by a foreign presence in one's felt-bodily presence, the metamorphosis of I into "me" does not yet say the same as "to me".

What we need to focus on is neither the reflective model (of the mirror, in the literal sense or not) nor that of recognition, as the former presupposes a previous familiarity with oneself and the second establishes an excessive dependence on others. Rather, we should focus on the model of involved and felt-bodily self-giveness.[3] We could also speak of self-affection, that is to say, the moment that integrates and unifies the otherwise normally "insular" situation of felt-bodily feeling: for example, when a newborn feels a sudden pain in their foot, which normally appears to them as nothing more than an external object of interest, they suddenly perceive the foot as their own. This feeling of "mineness" is instead completely absent in an anosognosic person, who perceives their foot as topologically foreign – just as a piece of paper or a pencil – and only recognizes it as their own abstractedly, for example by looking in the mirror, with a perfectly Cartesian reifying style.[4] After all, for perception to be not only a cognitively oriented, ocularcentric and distal perception, but generally "how one feels" (involved, engaged) in perceiving what one perceives, perception must also always entail co-perceiving (the quality of) one's own existence.

At first it could be objected that affective and felt-bodily involvement is not necessary to legitimize one's identity, as it suffices for one to coincide with one's qualitative attributes, but this would be a very dangerous move indeed. As shown by Hermann Schmitz's arguments, which I have partly mentioned (cf. 1.4), in order to know that certain attributes or even thoughts are identical to me (including the Cartesian *cogito*) – relative identity – I must already assume I can identify myself as me, because I already have self-consciousness. If identity is reached by self-attribution, in short, there is always one crucial step missing: the one that proves that I am indeed the one identifying with the attributes, that I and not others (which is hypothetically possible)[5] have my various properties (the fact that I am a case of various genera, such as males, professors, the inhabitants of Rome, etc.). As Lichtenberg said to Descartes, instead of "'I think' we should use 'it thinks' as we would say 'it lightnings' or 'it is raining'".[6] To put it in more abstract terms: in order not to fall into a *regressus ad infinitum* of the self-attributing process, it must be said that reflective self-consciousness can initiate the necessary processes of subsumption under genera only because it presupposes a previous and unconscious familiarity with itself, a more primal reflectivity that, not being cognitive-attributive, must be conceived rather as an affective certainty. It is not knowledge, which by principle can be revised, but a case of *tua res agitur*!

The neo-phenomenologically qualifying point here is mainly the ontological, and not only epistemic, distinction between subjective facts and objective facts. That is, between the meanings that only one person can talk about (using their own name), and those that everyone can talk about, insofar as they know enough and have the necessary language skills. However, despite referring to the same person, objective facts do not contain what that person really is. As Schmitz puts it, "the subjective does not consist [...] in a position on the terrain of objective facts (relational subjectivity) but in a different kind of factuality: the factuality of subjective facts for someone" (Schmitz, 2005b, p. 6). This means that only subjective facts (or meanings) express what we really are and (when we do not ignore them, of course) make us truly self-conscious subjects. And this is true regardless of whether these subjective facts precede their neutralization into objective facts, or whether the latter represent rather a (healthy) moment of crisis. Only when involved can the subject say "I" with a surplus of meaning that, guaranteed by "me" and above all by "to me", is always affective and embodied and not banally a matter of content, as in objective facts.

Now, according to Schmitz, a person is undoubtedly a conscious subject endowed with the faculty of self-attribution. But a person can identify something with themselves, thus avoiding the fallacy of *regressus ad infinitum*, only thanks to a pre-personal self-awareness that needs neither identification nor self-attribution. In fact, this self-awareness is based on affective involvement or, if you prefer, on subjective facts, which are much more than simple properties of a self: the latter, in a way, appears to melt like salt in water (cf. Blume, 2003, p. 33). Most of my involuntary activities are not really mine; instead, a feeling that atmospherically attacks me from the outside and affects my felt body, notwithstanding its coercive and initially impersonal character,[7] is necessarily mine – a "first-person" fact, so to speak. As Husserl himself seems to have done at least on one occasion,[8] here one could use the term "mineness" or "me-ness" (Schneider, 1959, p. 100), meaning the mark of the facts that can only be described in the first person. These facts differ from objective and neutral ones not so much in their content but rather because I, and only I, can talk about them in my name. I repeat: despite being "subjective", these are fully "facts"[9] because they are more unquestionable and urgent (like pain, an emotion, etc.) than a merely cognitive fact. But let's try to follow Schmitz's line of reasoning.

> The possibility of being self-conscious without the identification of something with oneself is based on the facts of affective involvement, even in their mere factuality, having the mark for the conscious subject of being-for-him as subjective facts, regardless of their ascribable content. This is only possible if, in them, he, for whom they exist, is also found. He has to be given as identical to them without identification. The identity we have here is absolute identity, the opposite of difference, not yet the relative identity of something with something with which it is identified. The absolute identity of being this (and not that) cannot always be taken as a given […] In absolute identity requiring no identification, someone can find himself if what he encounters without leeway coincides with what happens to him. This happens in flinching, in strong embodied contractions, in the sudden dawning of something new, for instance in fright, in an overwhelming twitch of pain, in a heavy jolt or gust of wind, if one is hit on the head or loses the ground under one's feet. Then the five elements *here, now, being, this* and *I* are rigidly joined in experience while orientation breaks down so that no features for the identification of something with something are available in this or that respect. This event I call the *primitive present*. The subjectivity of the subjective facts of affective involvement is based on it, for in it he is found for whom these facts are subjective. But, as has been shown, the possibility of self-ascription, which would otherwise not have a relatum, is based on them. Thus, the person as a conscious subject with the capability of self-ascription is only possible by means of the primitive present.
>
> (Schmitz, 2019, p. 62)

Without going back to the neo-phenomenological theory of primitive presence/present, let me recall one point. The reflectivity of the I, which in its common meaning (saying, for example, "I am sad") already entails some distance from the affective situation, is only possible if it presupposes the previous unconscious reflexivity of the "to me". But then it is permissible to affirm that the "to me" presupposed by the person is both impersonal and radically subjective, a sort of ur-subject whose personal-self-attributive identity is a neutralization like the distinction of the other from the self (cf. Fuchs, 2000, p. 130). In this sense, *contra* the Indo-Germanic language, "the word 'me' [is] to be understood not so much as a pronoun but rather as an adverb (a bit like 'here' and 'now'), one that does not denominate a thing but characterizes a milieu, just as by the word 'here' one does not refer to a thing ('the here') but rather to what is here, in the immediate milieu" (Schmitz, 1994a, p. 15).

But how does the "to me" of absolute identity (pre-personal subjectivity) become the personal I in the common sense of the term? Through the development of the five sides (and of course especially that of the I) which, as we already know, according to Schmitz are originally fused in the primitive presence/present (here/space, now/time, being/non-being, this/relative identity, I/other). Thus neutralizing (also linguistically) the subjectivity of meanings – the internally diffuse meaningfulness of a situation – the person marks their access to the impersonality typical of neutral facts. This impersonality is made possible by personal emancipation, able to neutralize meanings and, at the same time, singularize the world (meaning both the conscious subject and the things with which they come into contact). But there's more, because at this level there is a further distinction, which does not exclude multiple intermediate grey areas: that between one's own personal world (the so-called personal situation, with which Schmitz globally replaces the psychic), gathering everything the person strives for or fights against, and the alien personal world, made by all those things that fade into anonymity and otherness.

A conscious and self-attributive person can always go back to the proto-identity life of primitive presence/present, as such foreign to any identification and reflection, by means of affective and felt-bodily involvement – as in the case of fear and pain, which for Schmitz are exemplary.[10] Only the collapse of the orientative dimensions opened up with personal emancipation due to a certain "felt-bodily communication" – in particular the one that "catastrophically" flows into laughter and crying but also the one established with feelings (especially if atmospheric) like shame and rage, desperation and satisfaction – would allow the person to have a "healthy" regression to the primitive presence/present and thereby achieve an integral re-subjectification. This "personal regression" waters down once again the distinction between what is one's own and what is the other's. While emancipation explicates, personal regression implicates it, that is, it goes from single meanings back to the diffused significance typical of holistic situations for which it is even inappropriate to speak of reality in the full sense.[11] Therefore, the person always oscillates[12] between personal emancipation and personal regression, either simultaneously or successively, based on whether they go one degree up or down along this fluid dynamic (the two states only have a relative and positional meaning). The person, understood as identification with something else, in between a full neutralization (de-personalization) and a re-subjectification (pre-personalization), is always reached too late. But what triggers personal regression (i.e. the pre-personal) is mainly the (at least initial) rational ungovernability of atmospheric sentiments – to observe them in a neutral way or, even worse, to govern them would be equivalent to not experiencing them at all.

As can be easily seen, for a neo-phenomenological atmospherology, engaging in philosophy means getting up from one's desk or moving away from the Cartesian masonry stove, so as to face the concreteness of ordinary life experiences. In so doing, we will immediately and "painfully" notice that it would be very unnatural to keep doubting what we feel as "irritations". We will realize that, just like the ancient Greeks, today we are still felt-bodily exposed, in an overall-protopathic or discrete-epicritical way (i.e. predominantly in this or that felt-bodily isle) to external, and therefore atmospheric, impulses and feelings, with respect to which our felt body acts as a sounding board, without wondering how it is able to access it or sensorially check its reliability. The impersonality of atmospheres thus also resounds in relatively impersonal parts like the felt-bodily isles of the lived body. This constantly reminds the person that they might turn out not to be the owner of the house which they erroneously believe to be their own, so to speak. The repeatedly recalled (also atmospherically induced) felt-bodily competition between tightness and wideness is therefore not necessarily already personal. Infants and animals, in fact, are emotionally affected by it and are aware of it even though they are not properly persons (Schmitz, 1996, p. 26), as their subjectivity is simply a tonalized *milieu* of states of affairs, programmes and problems whose meaningfulness is still only internally (chaotically) diffused. When one laughs, for example, what one laughs about belongs to the level of regression, while the laughing subject pertains to the emancipated level, to which one can always return from the regression level. Similarly, when we speak, we follow a linguistic routine that pertains to the primitive presence/present in its fluidity, but whose content depends on the singularization and explication of the unfolded presence/present.

But as the "bearer of the zero point" of absolute subjectivity, the felt body is something pre-personal. Therefore, it can be said that the neo-phenomenological crusade for the externalization-impersonalization of affectivity goes hand in hand – in only apparent contradiction – with emphasis on the salience of subjectivity and, if you will, of personality, understood in an absolute way and largely unrelated to the common notion of "person". In this sense, subjectivity is no longer the land to which the *reductionist razor* exiles all that is too vague or complex, but the sphere of extremely precise subjective situations. Consequently, the person, no matter what their "style" is,[13] is a conscious subject that has an ascribable content and can singularise and explicate the internally and holistically diffuse meaningfulness of situations they encounter only if they have not completely emancipated themselves from (and can still access) the primitive presence/present. That is, only if the person is always "able" to appropriately fluctuate between the poles of the primitive (implicating) presence/present and the unfolded (explicating) one. Since "human life is usually distinguished by a varied swinging as well as an ambivalent shimmering between the two forms of present" (Schmitz, 1968, p. 4), and, as a consequence, the unfolded present is nothing but a labile and "fortunately" never definitely acquired stage (cf. Schmitz, 2015, pp. 119–137), our image of the person must change completely: it appears to be a chaotic phenomenon, infinitely (and therefore weakly) undecided about personality or pre-personality,[14] that is, about the five polarities and therefore also between primitive and unfolded presence/present, which also means between subjective (atmospheric) facts endowed with binding validity[15] and neutral objective facts.

According to this approach, for which the impersonal or pre-personal is a background acting with extreme precision in each (erroneously underestimated) routine, impersonality and absolute (not just positional) subjectivity paradoxically coincide.[16] Perhaps we could speak of a depersonalizing re-subjectification, whose oscillation between regression (mineness) and emancipation (personality) circumvents both the risk of a too solid introjective subjectivity and that of a metaphysical and all-catching impersonality. Paying attention to this absolutely subjective pre-personality does not mean disregarding the experience of everyday life for the benefit of some suggestive but too "sedentary" hypothesis along the lines of "too bad for facts!" The lucid awareness of the fact that in our life many things are not at all ours does not involve any disinterest in what *can* be ours and with which we hope to contribute to the anonymously sedimented "sense" that we encounter in our being-in-the-world. To use a literary quote:

> I mean, isn't that the most cherished pre-posthumous dream of all? The news of our premature demise catching everyone so unprepared that beautiful women have to leave fancy dinner parties to be alone for a while, their poor husbands looking around confused; grown men find they can't finish their after-lunch remarks at the Founders Club because they're so moved. Children wake up sobbing. Dogs howl, hounds begin to bark. All because something essential and ineffable has been erased, and the world knows it and can't be consoled.
>
> (Ford, 2006, p. 53)

Too melodramatic? Too metaphysically subjectivistic? Maybe, but perhaps not entirely inadequate to the existential commitment of all of us. Including those who, while sitting quietly at their desk, claim they prescind from the person and address the impersonal, except in real life they actually monumentalize themselves, cultivating obsessively (and perhaps emphatically writing) their own biography.

Notes

1 Here I am partly following Wiesing (2014).
2 For a presentation and a (partial) criticism of this "oblique thinking" (Böhme, 2012, pp. 188–193), cf. Hoffmann (1998).
3 That is, on pain, felt-bodily constitution, biography, language, but inexplicably (Böhme, 2012, p. 37) not on mentality, which perhaps he considered too disembodied.
4 Fuchs (2000, p. 98 f.).
5 "I must already know that I am the one who is signified in order to have a reference for my self-attribution" (Schmitz, 2015, p. 175).
6 And there is obviously no subject placed before and behind the event in these quasi-thingly dimensions (cf. Griffero, 2017a).
7 It is a holistic situation, which cannot be reduced to the cause of a single object. Indeed (as in the case of wind) it is only perceived in its effects (Ogawa, 2001, pp. 117–119). Some refer to it as a sort of pansphere (one of the meanings of the Chinese-Japanese term *ki*) in which world and I are perfectly blended together. Cf. also Hisayama (2014, p. 42 ff.).
8 According to Landgrebe (1974, p. 478): "the discovery of what is mine precedes the discovery of I".
9 The notion of subjective fact only sounds like a *contradictio in adjecto* within a mindset that is totally colonized by the natural sciences.
10 "Without receptivity to fright people would be unable to identify something as themselves" (Schmitz, 2005a, p. 13). For one to be a felt body means that one can be cornered and scared (Schmitz, 1989, p. 219).
11 "Reality appears all of a sudden, breaking the duration of subsistence in primitive present/presence (for example, when a sudden and intense noise wakes one up from dozing)" (Schmitz, 2005a, p. 29).
12 Hence a latent hysteria in humans as such (Schmitz, 1997, p. 173; see Stepath, 2006, p. 124).
13 For example: pride, irony, stoic imperturbability, sober realism, etc. (Schmitz, 1990, p. 155).
14 It is the logic of endless undecidedness, to which Schmitz recurs several times to avoid contradiction (cf. Blume, 2003, pp. 48–51). For this logical criterion, see Schmitz (2008a, p. 115 ff.; 2013).
15 For binding validity in law and religion cf. Schmitz (1973, 1977, 2012c).
16 This doubling up of impersonality happens differently from the way foreseen, in a slightly different sense, by Merleau-Ponty. For him, "each one of us must be both anonymous in the sense of absolutely individual, and anonymous in the sense of absolutely general. Our being in the world is the concrete bearer of this double anonymity" (Merleau-Ponty, 2002, p. 521).

7 Condemned to sense (and expression)

Eight theses on atmospheres as expressive wholes

Some, invoking conceptual "nudism", accuse any philosophy of expression of being marked by a "pathetic fallacy" and of giving in to some form of aestheticization, anthropomorphism and even animism (forms which are automatically stigmatized as regressive in any context that is even vaguely Cartesian). Well, the approach I am outlining here rejects the core idea of such "nudism", namely that an entity (a thing, an environment, a property) is perceived as autarchic and fully autonomous rather than for its expressive qualities. Instead, I believe that entities are first of all *aistheta*, something "expressive" able to exert some "impression" on the perceiver (not only visually). Indeed, a foreign language never appears as a simple set of physical sounds but rather as a semantic manifestation (exactly as the mother tongue appears to the child in the prelinguistic phase), otherwise one would not be able to even consider it a language and would not be aware of not understanding it.[1] Similarly, from a non-reductionist but *lebensweltlich* perspective, the world is never a set of sense-data that we "first" introject and "then" attribute a projective (subjective and/or historical-cultural) semantic evaluation. We would not evaluate an image, an external quality, if we hadn't already "undergone" its character or expression,[2] if you will. This is because "there are no experiences devoid of sentiment just as there are no completely uniconic sentiments" (Klages, 1964, p. 423).

However, it is a bit surprising that today there is again talk about expression. In fact, since expressive qualities were replaced with simple (mainly statistic) numbers even in the humanities, the so-called age of the "science of expression" – whose peak can be roughly placed between the two world wars – has appeared to be a thing of the past. The residual interest in expression focused no longer on how to understand the expression of others (if not of the whole world), but on how to optimize one's own expression for personal or collective success. The millennial (theoretical and even metaphysical) physiognomy soon became a prêt-à-porter knowledge aimed at optimizing one's professional performance. But things, obviously, have changed. Despite its irreducible semantic ambiguity and its difficult conceptualization, further undermined by the repeatedly mentioned relativity of every linguistic categorization of the emotional, expression seems to be once again indispensable for a phenomenology that focuses not on strictly cognitive processes, but on the "pathic" perception of what happens (to us) in our surroundings. Of course, theoretical needs are undoubtedly affected by often unspeakable pragmatic and political requirements: for example, the great interest in expression of the early 20th century, as known, was motivated by military applications of psychology. However, the needs of today's influential biometric security or occupational research will probably affect quantitative and analytical investigations of expression but not aesthetic-phenomenological ones.

In any case, the neo-phenomenological-atmospherological approach (Griffero, 2014a) outlined in the previous pages necessarily runs through the combined effects of the strictly dualistic couplet introjection-projection, blatantly inadequate to describe the phenomena of our everyday experiences. What we must reject is the idea that, just because we are first and foremost sure of ourselves, we can then (laboriously) infer something about others, whose experiences would be as inaccessible as a black box. My proposal for a pathic, quasi-thingly and atmospherological aesthetics (cf. Griffero, 2014a, 2016a, 2017a) rests on overcoming dualism, which is presupposed by the linear and centrifugal metaphoricity typical of the intentionalist paradigm, and by the idea that access to the world necessarily requires some medium. I would now like to posit a few simple theses on the expressive nature of atmospheres and, at the same time, on the atmospheric nature of expressive qualities.

Thesis number 1: atmospheric expressiveness is an irreducible qualitative whole

As feelings effused in the surrounding lived space, atmospheres are protypically resistant to the perceiver's attempts of projective transformation. If this were not the case, one would only perceive the atmospheric expressiveness favourable to (and therefore understandable for) one's subjective mood – a circumstance that is continually denied by everyday experience. In their relative cognitive impenetrability, due to the fact that they are first pre-dualistic and pre-reflective impressions able to "infect" bystanders, atmospheres can also be explained as expressive force fields. An atmosphere, in fact, is a whole of tertiary pericorporeal qualities that, as "emergent" superstructures filled with semantic-axiological qualities, impose a corresponding pathic condition on the perceiver thanks to their authority (and not only authoritativeness). The widespread submissiveness spread by a charismatic person on their interlocutors or the abandonment to which the spectator is induced by a sublime cosmic landscape are anything but illusory qualities. Because of their relative repeatability, inter-observability and even, within certain limits, crypto-normativity – which makes them reliable reasons for our affective judgment of the world – atmospheres certainly rely on expressive qualities, whose intra-atmospheric role must now be clarified.

In short: are atmospheres the (perhaps supervenient) aggregates of various pathic expressive qualities, or the (more or less stable) effects of their hierarchization? Do these expressive qualities uniformly tonalize the atmospheric field, or do they preferably condense in some (perhaps surprising and unexpected) areas of the perceptive structure? According to Garrick, a good actor should be able to perfectly express their character even through their foot: isn't such foot atmospheric? At this point it is sufficient to speak of "negative conditions" in the relationship between atmospheric wholes and expressive components. Certain expressive qualities do not infallibly guarantee the expressiveness-atmosphericness of the whole: not every blue sky "smiles" at us, not every fog saddens us, etc. However, it is almost certain that their absence – the lack of certain motivating and stimulating quasi-objective patterns – fully prevents "that" atmosphere from arising (for the benefit of a different one or of an apparent non-atmosphericness). One thing is certain: atmospheres are holistic expressive *Gestalten* that somehow emerge from their expressive manifoldness.

Thesis number 2: atmospheric expressiveness is not given by discrete sense-data but by internally meaningful chaotic-manifold situations

The historical enemy of a philosophy of expressive autonomy, as known, is associationism (cf. thesis number 6). An even earlier enemy was atomism, according to which an expression would be "reducible" – no term was ever more exact – to a complex of sense-data or stimuli whose action would be empirically testable. But as Baensch (1923, p. 2) observed very early on, the atmosphere, which here I equate to an emotional tonality (*Stimmung*), "hovers over the landscape, fills it and pervades like the light in which it appears, like the scent it emanates; it is part of the overall impression and can only be separated from it as a particular component through abstraction". This is true for every single expression: Merleau-Ponty (2002, p. 293–296) showed this by underlining how the same face seen the "right way up" and seen "upside down" are indistinguishable for the thinking subject but completely different for the perceiving subject. In fact, even though the second face in principle has the same optical elements as the first, the perceiver doesn't even see it as a real face. And this is because they lack a direct "grip" upon their percepts, which provides them with their prepersonal perceptive prehistory, which, adopting a relatively different vocabulary, should be seen as a form of pre-reflective, holistic and felt-bodily communication with the world (cf. thesis number 7), whose synoptic sedimentation is one's lifeworld.

An expression is therefore perceived, first of all, not as a set of singularities but as a meaningful "situation" which, as we already know from New Phenomenology, is made up of states of affairs, programmes and problems, and can only be dissected into discrete components *ex post* and, in any case, in a very artificial way. No matter if one calls them expressive or, as Schmitz does, "impressive" situations: all impressions, in fact, are aroused by some expression and in turn trigger some expression (conscious or unconscious, real or virtual) in those who feel them, resonating in their felt body. This physiognomic and non-analytical approach, which goes back from the impression to the expression, cannot make any demarcation between what is allegedly fixed (the phenomenon, from a physicalistic standpoint) and what instead is changeable (its expressive value), because the appearance of the expressive phenomenon itself, even in its atmospheric form, always exhibits a complete fusion and consubstantiality of materiality and expressiveness.

As I have been saying from the very first pages of this book, far from distinguishing a perceptive-sensorial "before" and a projective-qualitative "after", the percipient immediately and "ecologically" captures an expressive property. The perceiver feels in their felt body, though not as if it came from it, the expression that generated the impression they are under. If you will, the perceiver has "a feel for" the irreplaceable "scent" of the encountered (word, colours, etc.), of the things that involuntarily "happen" (the latter being a key term in a pathic aesthetics). After all, if someone perceived simple physical data, devoid of any expressive-atmospheric value, they would probably suffer from a pathology of perception analogous to aphasia, where one uses the words correctly but ignoring their intimate sense relationship with the world.

These first observations already suggest that passing from a strictly physiological description – committed to explaining the perception of the expressive on the basis of sensory atomism (sense-data, even physico-chemical) and of "separatist sensualism" (focused on the distinct sensory channels) – to an axiological-affective description is inevitably a dangerous move in an analytical-reflective context. In fact, this context is obviously blind to tertiary qualities, denying their phenomenological-experiential priority, but also to primary qualities, which are experienced in a necessarily situational way unless one claims to "reduce" them on the quantitative level. In short, an analytical-reflective context is foreign to the present approach, so rigorously phenomenological as to exclude from its investigable data even the neural processes in charge of perception physiologically understood (after all, the brain is not a phenomenon!).

Thesis number 3: atmospheric expressiveness eludes the traditional stratified conception of the relationships between nature and culture

According to an "artificial" (not to say artificially experimental) research such as Paul Ekman's, notoriously founded on a 15-second observation of the photograph of a face, or as the research on the ontogenesis of emotions in the first three years of life, some emotions would be fundamental, universal and pre-cultural. From the adaptive perspective of evolutionary epistemology, the expression of these emotions seems to have basically two functions. On the one hand (inorganic function) that of optimizing the reception and processing of external stimuli and of modulating the lived feeling. On the other hand, that of expressing – also unwillingly – such emotions to one's conspecifics, with consequent important effects both on social integration and on mutual understanding. Unfortunately the artificial situation of such empirical studies is a complete betrayal of the lifewordly situation, which from my point of view is the only appropriate context (both generative and receptive) of expressiveness.[3]

It is reductionistic to consider expressive qualities as projective integrations posterior to presumed elementary physical data. However, one cannot uncritically embrace the usual stratigraphic alibi, by which there is a set of expressions universally used to express basic feelings (usually six or seven) and another set made up of complex expressive forms that are instead very different from a cultural but also historical and geographical standpoint. A concrete individual, outside of the psychologist's experimental lab, has an overall holistic expressive behaviour that is largely determined by the environment. The nature/culture distinction, which is inevitably presupposed by the differentiation between basic and complex expressions, does not apply here (provided that it ever does).

First of all, it is not obvious that it is really unnatural (less authentic?) to react expressively on the basis of social rules. But, above all, since only the contextual situation can help us understand which of the innumerable meanings of laughter we should attribute to our partner's laughter, there is no such thing as laughter perceived as devoid of expressive values. Are they laughing because they are happy to see us or because they are mocking us? Are they formally greeting us, or are they experiencing an involuntary muscular reflex? Are they bitterly smiling about what happened to them, or are they about to sadistically throw a sarcastic accusation at us? And so on. Not to mention that the stratigraphic conception of the relationship between natural expressions and cultural expressions disclaims the already mentioned fact that, like feelings, expressions are rarely discrete and univocal phenomena. Indeed, they are mainly situations (in a meaningful sense), and maybe even mixed states – which, precisely for this reason, are more effective in the long term – in which, for example, the impression suggested by something that, while scaring us, also attracts us (like the sublime) is expressed in a necessarily vague way.

Thesis number 4: atmospheric expressiveness can be both intentional and unintentional

Before discussing atmospheres as expressive wholes, it is fundamental to make a premise: an atmosphere can be hardly explained as a product of organic beings,[4] let alone as something intentional. Therefore, the atmospherological approach must stay out of the thorny debate on whether the subject's experience coincides or not with its external expression. In a way, atmospherology can legitimately adopt, albeit in a rather different philosophical contest, the behaviorist thesis, which in fact is anti-dualistic and anti-associationistic.[5] According to this view, instead of manifesting something internal, expression rather expresses (cf. thesis number 5) a relation between the living and the world, without one being causally prior to the other, and without the expression being willingly manifested "for" someone. This position is quite distant from Husserl's, by which expressions are only speeches that, unlike gestures and facial expressions, have an explicit communicative intent and, precisely for this reason, do properly have a meaning (cf. Husserl, 2001, II, pp. 187–188). Even sticking to the noble and traditional rhetorical conception of expression, it is easy to imagine that even a totally involuntary expression can be rhetorically influent on a certain target.

In my case, there is no need to resort to vaguely animistic hypotheses given that, in its atmospheric value, expression shows a certain relation – not perceivable as such – between a section of the (lived) space and other intramundane situations, including the perceiver's. Of course the mountain peak does not worry about the approaching climbers. However, this portion of nature – by the virtue of some of its immanent Gestalt qualities and of the felt-bodily condition of the perceiver – might express for them a precise affective state, which can only be considered the outcome of subjective projection from a very superficial perspective. In short, just as the frowning face of a man with a grim look "expresses" something threatening like storm clouds, so "a violin can moan and exult like a man" (Schmitz, 1989, p. 184). When looking at an acrobat on a tightrope and feeling like we are with them, their perceived affective-expressive movements are mirrored by corresponding feelings in the observer (this is Theodor Lipps' known example). The observer in this situation feels no empathy, motor imitation or sympathy (ibid., p. 188), but rather a sort of "fascination" due to a felt-bodily communication (cf. thesis number 7) that bridges every gap and does not presuppose any psychism in the expressive source (indeed, the spectator could also be fascinated by a football during a match!).

Thesis number 5: atmospheric expressiveness is not the external manifestation of something internal

This is a truly crucial point. It is obviously difficult to avoid the idea, almost irresistible in certain languages (think of the German *Ausdruck*), that what expresses itself outside (*aus*) is an initial impression (*Eindruck*) on the inside (*ein*). However, my assumption is quite different: everything (animated or not), insofar as it appears, "watches" us – this is my interpretation of Merleau-Ponty's "everything has meaning" – and therefore expressively addresses us. Let's leave aside extreme cases, like the asemantic condition by which the alexithymic feels nothing about their surroundings, or like the pan-expressivism that induces the paranoid and the schizophrenic to believe that everything expresses something that "concerns" them. In a non-pathological situation of atmospheric expressivity, in fact, more than with discrete expressions (where they exist), one can certainly exclude that expressing always involves a psychic or, even worse, intentional activity that the competent interpreter should retrace. The modern use of the term "expression" in the aesthetic (musical and especially pictorial) field is largely indebted to the 18th century overcoming of the mimetic principle in favour of the "pathetic". It should now be extended to the point of ruling out the idea that expression ever imitates anything at all – not only nature and/or clear and distinct ideas, but also the interiority of a subject. This is the step forward I propose.

As known, the Romantic and Enlightenment overcoming of the subject's mimesis took place by seeing expression and especially art – roughly identified with beauty – as the finite manifestation of something infinite (the absolute). And this was in open antithesis both with the 18th century aesthetics largely influenced by Pietism, and with psychological aesthetics, which would eventually impose the commonplace view (based on music and poetry) that expression is the physical extrinsecation of some inner spirituality, both individual and collective (think of Spengler's physiognomy of cultures). The overcoming I propose, instead, simply considers expressive, for a perceiver of course, any form that can be perceived. In short: a (neo) phenomenology of the expressive does not have the slightest need to consider expression as the external manifestation of an internal spirit, be it subjective-individual-biographical or objective-collective-historical. In fact, for this approach there is no inner space that, in its relative mysteriousness, awaits to be "interpreted" and "understood" by sufficiently congenial interpreters.

Therefore it is not necessary to suppose an internal psychism behind every expression. If, following Merleau-Ponty, expression has a meaning, it is because the expression of anger, for example, "is" anger itself, its incarnation (cf. Waldenfels, 2000, p. 224) – clenched fists, frowning face, flushing, etc. This expression is neither simply the outward symptom of a prior internal anger that caused the outburst and which would, so to speak, be a model for different expressive situations, nor a third-person information about more or less objective facts that are independent from it. Nor is it a sign that can be traced back to a code by means of naturalistic translatability rules, and that for this reason, perhaps, can be transferred with no losses from one medium to another, like "information".[6] Sure, one can fake it, but the anger of the stage character – Stanislavski permitting! – cannot possibly be equal to real anger, not even if the former is the result not only of the role but also of an authentic empathic as-if. As mentioned, no expression is such *per se*, in the absence of some reception: it only becomes expression if it is "for someone". However, it is certainly far from always being the expression "of someone" – let alone with this someone being its conscious author.

This way we can disprove not only trivial psychologism, but also the early 20th century expressionist model, which followed a "dark inward path to the layers of creativity, to primal images, to myths, and in the midst of this horrorous chaos of disintegration and inversion of values, so as to arrive, by force, by law, by serious means, to a new image of humankind" (Gottfried Benn in 1914; cit. in Schürer, 2005, p. 247). It may very well have been the foundation of an abstract and anti-bourgeois world, but this artistic expressionism never managed to get rid of interiorism either: it merely refined and stylized it, disdainful of every social "blah blah" and of the vilified petty-bourgeois intimism. Conceiving expression as what "compels stringency in what it expresses" (Adorno, 2004, p. 18) also fails to get rid of interiorism and rather perfects it, removing it from immediacy (which is biasedly considered "bad"). Now consider the attempt at tracing expressions back to their signalling-communicative social function (the so-called audience effect, cf. Fridlund, 1994, pp. 145–168): one might say, for example, that a sad face does not so much express outwardly an internal psychic state, but rather signals the need for help to the communicative partner. Well, even in this case there is still an internal element – the need for help – and, on top of that, expressiveness is deprived of any adaptive value: in fact, unless we assume that the predator wouldn't take advantage of it, the prey would incite the former to attack it by blatantly expressing its weakness.

The principle of extrinsecation of the internal is also called into question by reflections that are not strictly (neo)phenomenological. First of all, even if I do not deal with this here, it is clear that there are emotions with no mimic, or at least visible, counterpart (i.e. with no corresponding external element). Secondly, the starting point of much (pre-Gadamer) hermeneutics is the so-called Schiller's law, according to which one can understand an expression only insofar as one traces back the lived experiences that generated it, and this principle is starting to wobble in any hermeneutics rejecting the dogma of comprehension as inversion of the genetic path. Even Darwinian studies, after all, question the perfect correspondence between inner emotion and external expression. In fact, to give some examples, it is possible to cry for joy (so there is an inner counterpart, but not the expected one), as well as to remain totally inexpressive in the face of intolerable pain (there is an interior that, paradoxically, expresses itself but exactly in the absence of expression), or even laugh, as happens especially in the East, to disguise one's negative emotions. Not to mention the so-called poker face, which by definition provides no information about the player's emotional state. Furthermore, there are psychic disorders where there is a huge gap between one's emotion and its potential expression, which can be absent (Parkinson's) or inappropriate (pathological cry and laughter, devoid of adequate stimuli).

And yet, despite these and other possible counterarguments, the principle that something internal is expressed on the outside – fatally applied also to atmospheric expressiveness – dies hard. Ludwig Klages, a fundamental author to the phenomenological science of expression (of the phenomenon as appearance) is not exempt from the internal/external dualism either. Without going into detail (cf. Griffero, 2015), it is easy to note that his opposition to any empathic, associationist and metaphorical explanation of expressive perception is rooted in a metaphysics of the soulful – "the felt body is the manifestation of the soul, the soul is the meaning of the living felt body" (Klages 1964, p. 163) – and therefore can never adhere to the externalism I am advocating here. Yet, strictly speaking, there is no reason to question psychic interiority for the thesis that, by means of an "expressive image, the vital wave transmigrates on the receptor of the image" (ibid., p. 173), understood as a "pathic" – extraspiritual (as much as possible) and not banally perceptual – contemplator of characters, primordial images and "essences" (non-objectifiable because they are in perennial becoming). The images that one pathically adheres to can actually be thought of as daemonic powers, as phenomena-appearances (in Klages' meaningful sense), even while excluding their soulful interiority.

To give just one example: "at the basis of every understanding of the line there is a lived experience of movement [by the perceiver], as some linguistic expressions still testify now: the road 'proceeds', 'turns', 'goes up', 'goes down', 'winds'; the spiral 'turns', the vine 'twists', the edge of the rock 'plummets', two lines 'meet' or 'run parallel' etc." (Klages, 1929–32, pp. 51–52). And this does not mean that these things (lines, roads, etc.) have a soul. Also, Klages (1964, p. 187 ff., p. 214 ff.) limits the expressive to involuntary processes, influenced more by life than by an extrinsic finality. However, given that the will (i.e. the Spirit) inhibits life, one can certainly assume that essences and primordial images are expressive independently of consciousness – of which one ought to "erotically" preserve the distance by virtue of supra-perceptual contemplation – even if they are devoid of life and therefore of soul. I will analyse no further the analogies and differences between atmospheres (in my sense) and Klages' images. Based also on the sober and probabilistic concessions of a certain Gestaltism (cf., e.g., Metzger, 1941, p. 97), I will only "inflationarily" claim that there are prototypical, derivative-relational and even spurious expressive-atmospheric properties. The latter, incidentally, are no less "true" than the others, given that the strictly phenomenological level of appearance exceeds the criteria of truth or falsehood (cf. Chapter 5): they are simply less representative of the external nature of affectivity that is the leitmotiv of my work.

Ultimately, the phenomenological approach circumvents the internal/external dualism only when, stopping at the perceiver's pathic and neo-phenomenological perspective, one has the courage to affirm that atmospheric expressivity does not manifest, but generates, a lived experience. Hence an externalization that, without identifying with the behaviourist one, investigates the "impressive potential in the presence of a possible spectator" (Böhme, 2002, p. 151). And it does so without degrading expressive appearance to a mere symptom, to a trace in which "to read" or worse "decipher" (noble metaphors, but far from philosophically neutral) something else, as if expressivity were a cryptic message waiting to be revealed. Thus eliminating the centuries-old compulsion to interpret the sign universe – a compulsion that the objectivistic disenchantment of the world has entirely exiled in the humanities – we should radicalize Goethe's sentence according to which "there is no inside, there is no outside, because the inside is also outside".

We would thus get rid of the distinction itself and conceive expressive phenomena a bit like statues: they are highly expressive and in some sense even animated,[7] but what counts is only their appearance (in other words, their actuality rather than their reality), and it would be absurd to look for something behind or within them. In short, what interests me here about the atmosphere as an expressive whole is only the fact that it is "the articulated way in which something breaks forth in the space of one's own presence", "a character in appearance" (Böhme, 2001, p. 168, p. 166). It is charged with its own potential for effect: this potential, of course, is sometimes also mediated by retentions, protensions and allusions to an imagined absentee, as in the case of deixis on the phantasma or imagination-oriented deixis (cf. Bühler, 1933), but is felt as immanent even in those essences or absolute qualities – that is, radically extra-thingly because unrelated to objects and situations – that can be perceived by excorporation (*Ausleibung*) according to Schmitz (1990, p. 153, following Conrad-Martius).

Now, it is clear that every expression "can" also refer to something other than itself. But in the name of their hermeneutic and then semiotic infatuation, 20th-century thinkers have long felt legitimized to unilaterally investigate this aspect. For example, because of its extraverbal specificity and individuality, the human voice evidently always provides also precise indications about the speaker (age, education, social background, geographical origin, physical and emotional condition, reliability, authoritativeness, etc.). Sometimes these indications are misleading, as for example in the case of a physiologically husky voice mistaken for a sensual voice; other times they are fictitious, as with an able simulator. In any case, duly contextualized in their being more symbols than unequivocal signals, these indications are obviously very informative and suggest a certain felt-bodily mood and attitude to the perceiver. They generate an overall expressive atmosphere, which, however, should be traced back not so much to an identifiable inner emotion but rather to an actual situational context. In some cases, the latter in itself is able to manifest how the speaker "feels" at that precise moment (Meuter, 2010, pp. 52–54), even if they are themselves unaware of it; in other cases, it can manifest the immanent affective quality of an inanimate state of affairs, "apparent in its mere appearance", so to speak.

Thesis number 6: atmospheric expressiveness cannot be explained in associationistic terms

For centuries (at least), despite undoubtedly appearing external to us in our daily perception, affective qualities have been explained on the basis of the already criticized paradigm of "access", by referring to crypto-anthropomorphizations and crypto-subjectifications of projective origin. These explanations are based on two assumptions: a) that a single datum is always (spatially, temporally and causally) associated with another single datum; b) that a situation is felt and evaluated in analogy with other situations. Habit and analogy: this is the "mantra" of the empirical theory of expression whose validity I intend to question here. But let's start with some observations.

First of all, this presumed external projection, condensed into an objective percept, of irreducibly subjective processes (psychological and/or chemical-physical) is supposed to apply to sense-data – thus transforming subjectivity (this wall is hard for me) in intersubjectivity (hardness for everyone) – but not to feelings nor, *a fortiori*, to thoughts. Why on earth should this be the case? Furthermore, given the alleged projective nature of "reality", how can we still call "external", as physicalism does, the nature of the stimuli acting on the sensory organs (cf. Klages, 1964, p. 377)? And, again, why is it that an expressive quality is more often transferred from nature onto us than from us onto nature? This is shown by the example of the weeping willow, which does not express sadness because it looks like a sad person, but because the shape, direction, and flexibility of its branches convey an expression of passive hanging, and *therefore* the willow suggests a comparison with the structurally similar psychological pattern of sadness in humans (cf. Arnheim, 1966, p. 64).

To make another example, the expressive difference of "cheese" and "Käse" shows all the inadequacy of the criterion of arbitrariness compared to a linguistic sensitivity that, at least within one's mother tongue, is instead always holistic, motor and synaesthetic, i.e. motivated and felt-bodily (and in this sense also atmospheric). This inadequacy applies even more to the perception of every expressive form. Therefore, in the preanalytical lifeworld, we are all naive realists, no matter how illusory this view is.[8] For example, certain mountains objectively appear "sad" or "threatening" to us and not just "dark"; a meadow is not only green but also "joyful" if illuminated by the sun, or "melancholic" in autumn or in cloudy weather. And these qualities are not simply deictic, as they do not refer exclusively to the projective mood of the percipient.

I have already analysed the neo-phenomenological reconstruction of the two-billennia-long introjectivist-psychologistic error, which is the origin of many mistaken theories of expression (Chapter 2 and Chapter 3). Instead, I claim that expressive qualities, and atmospheres as their aggregates, are certainly not subjective only because they require a subject in which to resonate – if that were the case, any knowledge obtained by our species would necessarily be subjective, too. But what I want to focus on now is the associationist explanation. Implying the projectivistic hypothesis, according to which the world has qualities only because it has been (unconsciously) filled with them by the subject, associationism is the heart of a real "cultural trash [...] full of associations, past experience, conscious or unconscious inferences, probability calculations, object hypotheses and projections" (Bozzi, 1998, p. 105). Despite often being a culturalist *petitio principii*, the associationist pseudo-explanation is hard to eradicate and may appear obvious, but it is not obvious at all, as I will now show through four passages.

a) First of all: "the perception of expression is much too immediate and compelling to be explainable merely as a product of learning" (Arnheim, 1974, p. 449). In neo-phenomenological terms, it is an impressive situation (an instant affective signification), be it latent and hardly perceived or blatant and aggressive in its quasi-thingly nature. Consider for example a "sensible working-class woman who sees at a glance a slight anger in her returning husband and a mute irritation in her son" (Klages, 1964, p. 379), or the buyer intuiting the cunning deceit of the seller: they both "feel" with absolute precision an expressive-atmospheric quality, albeit without being able to analytically and reflectively explain the link between physiognomic traits and their impressions (this is what the scientist of expression will do). Given that in a way we grasp expression the more precisely the less we try to infer it from discrete data, I believe the associationist process can be considered refuted. If, for the first time in our lives, someone tried to punch us, we would instinctively dodge them: not because we associate it to some previous experience, but because the motor suggestions and synaesthetic characters of the punch are "immediately" hostile. The analogical reference, at most, can be a sort of later verification of our innate and instantaneous expression-feeling link. "Being dreadful is a property of this Japanese mask, an inexhaustible and irreducible property which constitutes its very nature – and not the sum of our subjective reactions to a piece of sculptured wood" (Sartre, 2002, p. 383).

b) For example, according to an entire psychological tradition, we would "deduce" that someone is angry from their flushed cheeks – hence a rationalistic integration of the more elementary empirical associationism – because we also flush when we are angry. The analogical explanation thus entails that we do not "see" and "experience" the others' anger but rather "know" about it based on our own anger. This view is very fragile. First of all it erroneously presupposes that one perfectly knows one's own anger, which is a lived experience that is very different from its possible expressive forms. Even verifying this experience in the mirror is no guarantee, because in this conscious reflected image the anger would inevitably be partly deformed (if only because it has already passed). Also, different individuals manifest irreducibly different expressions. Therefore, we learn about an expression from the image of the other rather than from our own. This is true for animals (especially pets), who grasp, and certainly do not project, a certain irritation in the severe gaze of their master. And it is also true for newborns (already, apparently, from the tenth week of life): in fact, they sense the positivity of their mother's smile (smile as atmosphere) and get attuned with it long before they grasp its empirical details and long before seeing their own smile and being able to make inference or analogies (cf. Klages, 1964, p. 379 and Waldenfels, 2000, p. 243). Similarly, newborns do not see their own agitation in a deformed face but rather what triggers it (cf. Klages, 1964, p. 294). It follows that "the outward gaze does not have internal criteria to explain what is foreign on the model of what is one's own" (Waldenfels, 2000, p. 243), but uses a "circular process of resonance" and "affective harmony" (cf. respectively Spitz, 1960, p.

43 and Stern, 1985). This "being-moved-together-in-shared-way" (Klages, 1964, p. 294) obviously refers to animated *Gestalten*, whose lived experience can (and should) be shared by the percipient, but also, as I have repeatedly underlined, to inanimate *Gestalten*. In both cases, this process finds its condition of possibility in a mimetic proprioception of the perceived based not so much on one's impressions but rather on structures governing both one's "behaviour" and the affordances of things.

c) The explanation based on past experience does not work. Not only if, by embracing imaginal heraclitism *à la* Klages, the perceived qualities of expression are considered unrepeatable, but also if one supposes an associative retracing. In fact, either we refer to some "first" immediate qualitative perception, which is also inexplicable by association, or we have to trace any expressive perception back to previous experiences, thus generating an infinite regress that can ultimately be only solved by evoking some "world of ideas (or expressions)" (Bozzi, 1998, pp. 170–171). To be clear: it is of course possible to judge someone's expression *also* by habit,[9] but this is a "generic" and in some sense misleading perception, since, being extraneous to a real lived experience, it would grasp the "what" of an "expressive movement" rather than the "how", which is its essential character (as we know) because it is pathic.

d) Projective associationism presupposes what it should demonstrate (Waldenfels, 2000, pp. 242–243): to explain why a mood is projected on a certain thing and not on another, one is forced to admit that it is precisely that thing (and not another) that suggests this hypothetical projection, and therefore one should ask "whether the objects themselves do not, after all, demand such a transfer" (Heidegger, 1995, p. 63).

The projectivistic-associationist pseudo-explanation is fundamentally anachronistic (Waldenfels, 2000, p. 243) because it supposes a perception of the inanimate-neutral prior to that of the expressive, whereas the order of the factors appears to be exactly the opposite (Scheler, not coincidentally, speaks of progressive "disanimation"). As a final objection, I would like to quote Sartre's fascinating example of sliminess, which possesses us as soon as we believe we possess it. Well, the (symbolic, Sartre says) relation between sliminess as antivalue "and the sticky baseness of certain individuals" is only verisimilar if one grasps "baseness [...] in sliminess and sliminess in certain baseness", that is, only "if we come to life in a universe where feelings and acts are all charged with something material, have a substantial stuff, are really soft, dull, slimy, low, elevated, etc. and in which material substances have originally a psychic meaning which renders them repugnant, horrifying, alluring, etc. No explanation by projection or by analogy is acceptable here" (Sartre, 1969, p. 605). Note well: we are born – and not "learn" – in a world whose expressive qualities, prior to any distinction between internal (psychic) and external (physical), act as ways of being for every entity or event. Is this not also the meaning of Husserl's passive synthesis?

Thesis number 7: atmospheric expressiveness implies an articulate felt-bodily communication

The relationship between atmospheric expressiveness and its sounding board, i.e. the felt body, is very complex and cannot be analysed here.[10] Suffice it to note, to use an analogy between emotions and atmospheres, that personal emotions prepare for action the perceiver's physical body that expresses them (Ellgring, 2010), whereas in the case of atmospheric expressiveness it is rather the perceiver's lived felt body that resonates, being driven to action and, more often, to the contemplation of the expressive qualities it perceives. Also, an emotion sometimes undoubtedly also functions as an interface, mitigating the reaction to a certain stimulus, for example by transferring anger to one's face rather than turning it into action. In this sense, the only possible analogy is that in atmospheric perception the perceiver's persistent emotion is strengthened or otherwise modified, up to generating phenomena of emotional interaction that are fundamentally still unexplored. Indeed, there is a certain fear of discovering that what unites individuals exposed to the same environmental atmospheres may suggest a somewhat undemocratic determinism (geographical and then political).

But the most important aspect is another: namely the fact that our relationship with the expressive wholes that I call atmospheres is a communication that, in the wake of Merleau-Ponty (*intercorporéité*) but above all Schmitz (cf. especially 1989, 2011), I define felt-bodily (*leiblich*). Far from being a mere epiphenomenon or an integration of functional behaviour, this encorporation (or, at the opposite pole, excorporation), as we know, is so pathically engaging that it felt-bodily re-orientates those who experience it. And this happens not despite, but precisely thanks to, the fact that it is unaware. In short, to make an example, a steep cliff generates a proud atmosphere around the percipient by virtue of the same "felt-bodily existential" (Fuchs, 2000, p. 202) whose transmodal transversality makes it applicable in different regional ontologies. And, just like a gesture, a look, a way of walking, a dress, and so on, every perceived form expresses and embodies a meaning for us. Regardless of whether this felt-bodily communication, as Schmitz notes (e.g. Schmitz, 1978, pp. 31–32, p. 54), is complementary-syntonic or antagonistic-dissonant, etc.,[11] it always acts as an analogical "bridge" between the perceived and felt-bodily perception. Prescinding both from (presumed) elementary and organ-specific sense and from higher-order inferential and associative activities (cf. Metzger, 1941, p. 67), this peculiar form of communication is based on motor suggestions and synaesthetic characters. The latter reveal physiognomies that are also felt-bodily (Volke, 2007, p. 196 ff.) and isomorphically present both in the perceiver's felt body and in the felt body of the "surroundings": think of the linear-angular or rounded figures indicating epicritic or protopathic or even erotically suggestive tendencies, and so on (cf., e.g. Schmitz, 1966, pp. 63–69).

Ordinary language is certainly correct in attributing expressive qualities to things themselves rather than to the perceiver. Indeed, we say that the room is cozy and comfortable; that the court inspires fear and possibly guilt in the defendant; that the bank as such suggests (especially when there is no economic crisis!) solidity and reliability (these are the examples given by Adolf Loos); that that door and not another invites us to enter. And this is because such expressive qualities "resonate" precisely in the perceiver's felt body, finding fertile ground in it. Another characteristic of atmospheric expressiveness is the fact that the environment pervaded by it does not simply arouse impressions, but also provides a solid framework for affective states, some of which would even be impossible without certain environmental conditions (without atmospheres). Atmospheres are therefore affections extended in space, "tools for feeling" (Slaby, 2014, p. 43) if you will, that are mainly centrifugal and without which we would neither feel nor act as we do. In fact, sometimes it is only by virtue of the external atmospheric expressiveness that a vague mood is incorporated and clarified, acquiring a more stable and more "authoritative" form, able to suggest an axiological orientation (in a broad sense). The felt-bodily communication, then, is the condition of possibility of an authentic affective enactivism, from which follows a (non-physical) spatialization of our felt body. Because of this, the issue of the internal or external character of (emotional and atmospheric) expression becomes somewhat obsolete, being reduced, at most, to a question of relative and transitory predominance (ibid., p. 38), or, as we have already mentioned, to a question of prototypical or derivative (and even spurious) atmospheric expressiveness.

Thesis number 8: atmospheric expressiveness is the outcome of affordances or ecstasies (but not properties)

As we have seen, the goal is the radical de-subjectification and externalization of the concept of expression, referring it neither to a subject who, indeed, would be "expressed" in it, nor to the "things" that it would presentify, but to the appeal that it addresses to the perceiver. The physiognomic value of this appeal is an external appearance without an interior, a structure that "does not 'represent' or 'signify' the corresponding way-of-being", because "if it exists, there is also that way of being; the latter is not behind [the structure] but inside it" (Metzger, 1941, p. 62). This means, for example, that joy does not consist in an ineffable internal psychism, already fatally reduced by the anti-narcissistic observation that one's expressive repertoire is never truly original, but in the act of "jumping for joy" as such, as a felt-bodily expansive manifestation. In other words, the expressiveness we attribute to atmospheres, as Gestaltism posits for every expressive quality, is directly supported "by the constellation of observable events which identifies its position in space, i.e. depends on them and is located where they are" (Bozzi, 1998, p. 115). However, to amend Metzger's (1941, p. 60 ff.) pioneering tripartition, expressiveness is located not only in the ways of being of qualities (e.g. virility) but also in terms of skeletal structure (angularity) and of qualities of material (roughness).

Then, it would be better not to use the term "property", which is a predictable qualification of a substrate that cannot be predicated of other things, but rather to use the term "ecstasy" (Böhme, 2017b, pp. 37–54), meaning the specific way in which every form,[12] in a way, comes out of itself and tonalizes the pericorporeal space of the perceiver. In fact, unlike properties, expressive and atmospheric ecstasies are neither a) something accidental that the object has nor b) something that can only be thought of, without appearing in actual and felt-bodily form, nor c) something that exclusively indicates some usability. So, things come out of themselves expressively, always interacting with their context and possibly with other contiguous ecstatic irradiations, up to forming a specific atmospheric field: this is why, for example, a bridge can express lightness regardless of its actual specific weight, or why something blue can be "colouring and tinting all other things at the same time" (ibid., p. 51). What follows is an expressive quasi-objectivity that, as mentioned, also applies to primary qualities if one observes not only quantitative properties but also qualitative ones, like motor suggestions. This quasi-objectivity, after all, is also the basis of "style", which ultimately is nothing but a formal-creative objectification of the felt-bodily reaction to environmental expressive qualities, able to expressively arouse determinate impressions in the observer thanks to a specific felt-bodily communication.

I have often related this ecstatic expressiveness with the Gibsonian notion of affordance (cf. Griffero, 2014a, pp. 47–54). Here I will merely note that the notion of expressive-atmospheric affordances – partly anticipated in the 20th century by the notions of "vital meanings" (Kunz, 1966), "invitation characters" or "values" (Lewin, 1926), "vitality affects" (Stern, 1985, p. 54) and "ways-of-being" (Metzger, 1941) – serves to account for the affective, perceptive but non-quantifiable meaningfulness that constitutes the lifewordly first impression. This meaningfulness is made up of demands-invitations that are effective without being necessarily true or adaptively functional, but above all that, despite being obviously related to the perceiver's "size" (in the broader sense), are never their exclusive property (cf. Gibson, 1986). Like Gibson's strictly "ocular" affordance, the atmospheric affordance never fully prescinds from some perceptive-cognitive context, and certainly does not articulate the environment-organism reciprocity only pragmatically. In fact, for example, it can invite us to pure excorporating contemplation or to simple curious observation.

Also, the atmospheric affordance is not circumscribed to elementary appetitive vectors, because it usually suggests affections that are completely irreducible to the elementary attraction / repulsion binary system. This is the case with the sadness suggested by intolerable beauty, or with the already mentioned atmospheric vicarious shame, but also simply with emotional expressions within a range that can only be more or less something (an atmosphere can simply be more or less tense, more or less inviting, etc.). And sometimes the atmospheric affordance is just recognized-revealed (partly also affectively, of course), without being truly felt. In this case, the subject does not necessarily suffer from some frontal disease or from anosognosia: simply, their dissonant felt-bodily disposition prevents them from being deeply involved. As Moritz Geiger would say, the subject is thus induced to see a *Stimmung* as something exclusively external because of the distinction between an observing attitude and an absorbing attitude (Geiger, 1911, p. 27 f.).

In the realm of atmospheres there are, therefore, definitely forms endowed with an expression as immanent and immediate as the one attributed by Gestaltists to *takete* and *maluma*, but also – hence my preference for an "inflationary" approach – forms whose expressiveness is somewhat "mixed" or even barely perceptible. In short, recall the Usher house: its threatening and melancholy darkness expresses a painful atmosphere of "insufferable gloom", which exhales from its "decayed trees, and the gray wall, and the silent tarn" (Poe, 1839, pp. 291, 293), and cannot in any way be changed or even mitigated by the visitor's previous mood.[13] The Usher house is therefore certainly very different from the more famous proustian *madeleine*, whose extratemporal happiness does not at all lie in the biscuit but, as Proust himself rightly acknowledges, only in the narrator, when "a shiver ran through me" (Proust, 1992, p. 60). For example, I never had an auntie Léonie who used to dip a *madeleine* in a cup of tea on Sunday mornings in Combray, and therefore there is no "ecstasy" for me related to that cup of tea – let alone am I inspired to embark on something analogous to the "search of lost time".

Notes

1 As rightly noted by Waldenfels (2000, p. 220).
2 Or image (for Klages, 1964, p. 252).
3 Here I embrace the objection to the ecological validity of such researches (Meuter, 2010, pp. 36–37).
4 Despite excluding the inorganic origin of the atmospheric, even Tellenbach (1968, pp. 43–44) is then forced to admit that even the inorganic is atmospherogenic as soon as it enters into a relationship with the psychic (and when is this not the case?).
5 At least in the unorthodox meaning of behaviourism adopted by Karl Bühler (cf. Friedrich, 2010, pp. 100–101).
6 Cf. Schmitz (2010). News remain the same even in translation and the boiling of water is always perfectly equivalent (of course only on the physical level) to 100 degrees Celsius, while a smile and a certain enthusiasm largely lose their original expressive quality in, say, the two-dimensionality of a photograph or the fixation of writing (cf. Schmitz, 1989, p. 179).
7 I am here following Böhme (1995, p. 169).
8 As Volke (2010, p. 180) claims for his own linguistic community.
9 As Klages (1964, p. 293, fn. 36) also admits.
10 For an introduction cf. Griffero (2017b).
11 Its perfection increases the more it presupposes a constant feedback, as in the case of interaction (mutual but not symmetrical) between initiative and resonance that can be found in every couple, even in rider and horse or driver and vehicle.

12 The constitution of forms "is ek-static; this ecstasy finds its simplest expression in their spatiality [...] No form [...] is reduced to the image of its line. It emerges from it" (Maldiney, 1966, p. 218).

13 I have used Poe's story elsewhere as a leitmotiv for a synthetic analysis of atmospheric phenomenology: see Griffero (2018).

8 Come rain or come shine …

The (neo)phenomenological will-to-presentness

Pathic aesthetics, which focuses on the atmospheres expressed by ongoing situations, implicitly involves an intense valorization of the appearance of phenomena (as opposed to their essence and/or material constitution) and of their presence (as opposed to their being only thought or imagined). The temporal and spatial dimension of presence is precisely what I will now dwell on, in order to rescue it from the exile to which it was condemned in the 20th century, not so much by daily life but by philosophical-humanistic culture.

Consumed either by the future, by all sorts of goals and deadlines, or by the past – apparently the only guarantee that something is (was) a fact – we usually end up disregarding the present and presence. So, the least that can be said is that presence and the present have never had good press in the last century. Merely dealing with something or somebody without planning ahead or simply lingering in a place any longer than necessary already seem to run counter to the predominant everyday teleology. By acting in this way, however, "we fail to notice what we actually are: life", because "promoting presence-awareness is also a matter of the awareness of our own vitality" (Böhme, 2014, p. 151). Driven by the intent of rehabilitating presentification from a spatial and temporal viewpoint, this chapter is specifically aimed at: a) addressing Hans Ulrich Gumbrecht's recent *kulturkritisch* approach to presence; b) defending (Schmitz's) neo-phenomenological theory of the present as the principle of subjective identity; c) and providing a link between these two rehabilitation strategies through the pathic-aesthetic notion of felt-bodily presence. Although there are other possible approaches to presentness, here I shall simply ask what kind of new understanding about felt-bodily presence/present can be gained by combining Gumbrecht's approach to presence with Schmitz's ideas about the present.

Today, some see the revaluation of presence as a real paradigm shift. Although I am increasingly allergic to the serial production or even deluge of *soi-disant* new paradigms that appeared in the humanities over the 1970s and 1980s, whose outcome has paradoxically been a pseudo-obligation of permanent criticism at best and a relativistic *horror vacui* at worst, I nevertheless must admit that the previously mentioned "affective turn" undoubtedly seems to also imply a new way of understanding presence. The present has long been underestimated in the modern era, first of all because of the time deferral implied on the one hand by the Being-prophecy underlying utopian-eschatological thought, to which present is nothing but a painful long wait for the Totally Other, and on the other hand by the contra-present antiquarian nostalgia for what allegedly preceded an irreversible time of decline. This underestimation, though, also depends on the more recent (from Kant onwards) social, cultural or interpretivist constructivism, according to which presence, understood as an appearance that seems to satisfy only a naive conscience, would be nothing but the very complicated and incomplete outcome of a necessarily mediated (and today digitized) access to the world.

If we broaden our view, it may also be argued that this underestimation of presence is the inevitable result of today's increasing pressure towards coolness as a modern form of detachment from everything. According to Peter Sloterdijk, for example, modern cynical knowledge expresses itself in a "crooked smile" reflecting a metaphysical dualism (essence/appearance, inner/outer, etc.), in a gaze that "transfixes things it does not penetrate and to which it does not really grant existence. [It] lets things know that they do not exist as real objects for it, but only as phenomena and information [...] as if they already belonged to the past" (Sloterdijk, 1987, p. 146). Still, even the cynical intellectual is nonetheless surprised and "offended that the things return this gaze" (ibid.). Things cannot therefore be easily dismissed and this very "nuisance" proves that the cynic is still emotional at some level (a robot, in fact, is not offended by anything!) and in some way involved in something present. Hermann Schmitz's new phenomenological interpretation of Modernity as the age of ironism (Chapter 3 and Chapter 4), characterized by the ability to freely (and today technologically) withdraw from all points of view and therefore also to embrace them all (cf. Schmitz, 2010, pp. 111–126; 2019, pp. 56–57), seems at least partially disproven by the many ideological and emotionally contagious dreams of the 20th century, which was all but free of affective involvements.

This rehabilitation of presence/present, however, wishes neither to delete the past, as a large part of today's naive technological optimism does, nor to reduce it to its traces, as Derrida claims. His idea that events are nothing but their traces always reminds me (half in jest) of a tourist's bad choice to visit a place not to experience its presence but to anxiously produce the traces (photos, selfies) that prove that he has been there. On the other hand, I do not either intend to distrust the future as such, as Ludwig Klages claims on behalf of the eros of distance. However, it might help to try to think, like him, that only past and present truly exist and constitute real time, going against common sense (understood as a phenomenon following the Promethean ideal). In fact, while the past is undoubtedly real, the future (the "time-to-come") is nothing but a thought – it's only what might not happen – and therefore a void, a mere chimaera for deluded minds. For this reason "the 'eschatologies' and the 'apocalypses' [are] the most terrible expression of the madness called 'history'" (Klages, 1974, p. 435). Now, my approach is certainly metaphysically much less ambitious. I simply wish to point out that the question of presence/present is something "that, only a few years ago, must have looked so utterly naive that nobody dared to ask it" (Gumbrecht, 2014, p. 8). The modern and almost ubiquitous success of the hermeneutic *Weltanschauung*, obliging one to interpret every kind of experience and thus to unavoidably defer the presence/present's sense, thus made a fatal deal with historicist radicalism, aimed at questioning objectivity and degrading the present/presence into a mere elusive moment of transition, totally negligible if compared to the alleged active transformation of the world.

According to Gumbrecht, presence reminds us that "things inevitably stand at a distance from or in proximity to our bodies; whether they 'touch' us directly or not, they have substance" (Gumbrecht, 2014, p. ix). This means that we should reject all the necessary assumptions of the so-called meaning culture (cf. Gumbrecht, 2011; 2012a, pp. 190–209, pp. 244–245). Let's begin with a) the observer's ex-centric role: concerned with finding the appropriate distance in relation to a world understood as a mere world of objects, this role has inevitably led to subject/object dualism. Then, b) the subject moved away from medieval thought, found Cartesian dualism and especially the Protestant (only) symbolic interpretation of many Christian theologemes, so that Christ's body and blood became only "meanings", something that simply "stands for". This transformation of the mass into a mere commemoration of the past also helped turning the subject into an unbodied and genderless entity, which as a counterpart to the physical world appeared as the only producer of all acceptable knowledge about the world. In this process c) the distinction between material surface and deeper meanings – in its many variants, including the Heideggerian distinction between Being and what is "present-at-hand" or "ready-to-hand" – increasingly appeared to be something natural, so that the truth was necessarily identified with something behind the surface (or the signifier), beneath a "body" whose hidden truths and thoughts would be "expressed" only in an imperfect way. Lastly d) the subject, now disembodied and spiritual, seemed to realize itself if, and only if, it was constantly motivated by a drive towards change and transformation of what is "only" present (and can be therefore brushed off).

However, it is too naive to believe that this list of charges can easily defeat the modern hermeneutic credo (i.e. the unbridgeable gap between the surface of things and their conceptual depth) and replace it with "a cosmology of which humans consider themselves to be part" (Gumbrecht, 2003, p. 83). Since the "desire for 'full presence'", after all, "is a desire without the possibility of fulfillment" (Gumbrecht, 2014, pp. 4–5),[1] it is better to think of the everyday situation rather as the "productive tension" and oscillation between "presence effects and meaning effects" (Gumbrecht, 2003, p. 107). This very restriction urges us to understand the felt-bodily communication between perceiver and world (made up of things and quasi-things), or Gumbrecht's being "in sync with the things of the world" (ibid., 117), not too univocally, both in a Heideggerian (as unconcealment of Being) and in a ritualistic[2] or even eucharistic sense (cf. Gumbrecht, 2012b, p. 218). To me, what's clear is that the specific mission of aesthetics is to go against this meaning culture, based on a subject/object dualism, on a preconception of the invisible depth as the final truth, on the repression of sensible perception and the body (especially the lived one), and on the so-called psychological-reductionist-introjectionist paradigm. Since its modern origin, aesthetics in fact has not only questioned concepts (even more so if clear and distinct) but has especially highlighted the sensible particulars on which we should linger for a while,[3] thus fulfilling "an enhanced need – and an increased desire – for encounters with presence" (ibid., p. 7).

In short, in a pathic aesthetics, the present is no longer an intolerable Procrustean bed and presence, a real blind spot for our Cartesian and meaning culture, can assert its rights once again. Paradoxically, the present is claiming to lifewordly become (more) present again, thus abandoning our (especially modern) secularized perfectibilism. Nor is it by chance that our present no longer seems able to "secrete" future expectations. Indeed, the current crisis of (spatial, economic, etc.) expansion expectancy and the resulting threatening scenarios necessarily challenge any trivial teleologism and justify, if not our squeezing into the present, at least a different chronotope. This is based on the "replacement of the *only telos* by *many short-term ends*, perhaps on the substitution of the idea of 'perfectibility' by that of *humankind's preservation*; maybe replacing *spiritual man* (called to improvement) by *physical man* (interested in his own preservation)" (Gumbrecht, 2012a, p. 24, original emphasis).

Regardless of Gumbrecht's (ibid., pp. 214–215) models of the sportsman or opera singer, a neo-phenomenological rehabilitation of presence/present focuses, more generally, on appearances or signifiers not necessarily coupled with a meaning: physiognomies without interiority (cf. Chapter 7, where I spoke of expressions without a corresponding interior). It may be useful, in fact, to philosophically widen Gumbrecht's ontology of literature, based on "reading for *Stimmung*"[4] and immediacy – "for what affects us in the act of reading involves the present of the past in substance – not a sign of the past or its representation" (ibid., 14).[5] This means placing re-presentation above representation (which has been predominant ever since Descartes), but also supplementing this approach with an aesthetics of atmospheres (as quasi-things) and a phenomenology of the felt body as their precise sounding board. I can only endorse Gumbrecht's view that "yearning for atmosphere and mood is a yearning for presence" (ibid., 20), thus replacing meanings with (even painful) present "moments of intensity", with atmospheres and moods. The latter do no longer need to be "deciphered" (ibid., 18) but have to be felt-bodily lived as fascinating and almost gestural sources of energy whose authoritative relevance we can neither arbitrarily generate nor make last longer (Gumbrecht, 2003, p. 99, p. 103). However, *contra* Gumbrecht, this doesn't imply a special "insularity", namely an aesthetic (literary) and for this reason too elitist "specific disposition". This, in fact, would seriously underestimate the epiphanic potential of everyday life experience, as in the usual coming and going of atmospheric feelings: they (often) come out of nothing, have a spatial dimension, are unforeseeable events, are fascinating thanks to their aggressive (quasi-thingly) authority, and so on.

In short: the rehabilitation of presence/present means avoiding both the indefinite and therefore frustrating search for a transcendental foundation and the necessarily relativistic constructionist approach, thus reintegrating three underestimated dimensions: space against (overestimated) time, appearance as such against any alleged substrate and as self-reference against something other, but especially the lived or felt body, which we don't *have* but rather *are*. Even an ontologist of literature like Gumbrecht recognises that "we are fast losing the ability to 'be' a body, that is, the ability to let the body be an enhancing condition of our existence" (Gumbrecht, 2014, p. 70). For this reason, it's only a small step, so to speak, from Gumbrecht's historic-cultural will-to-presence – the "yearning for moments of presence in our broad present", the need "to grasp them and to be open to their fullness" (ibid., 79) – to Hermann Schmitz's ontologic-neophenomenological will-to-present: for me, these are two of today's most relevant and solid declinations of the Heideggerian being-in-the-world.

Like the importance of the presence, the importance of the present was also strongly reduced in 20th century philosophy. For example, Bergson gives priority to the past, as the present is something elusive compared to the *durée pure*. Heidegger instead gives priority to the future as the only source of authentic temporality. And Husserl makes the present too thin, taking the primordial impression (*Ur-Impression*) as a horizon that always includes retention and protention. In contrast with this disregard of the present and, of course, with the tendency to divide time as if it were an extended object, the present is for me the "now": an element of the time-quality that is immediately evident despite any objection of discursive thought (from ancient scepticism onward). In this sense, the present must have "nothing to do either with the smallest part of measurable time that we are now in the process of representing to ourselves or perceiving or, *a fortiori*, with the infinitesimal of physical theory. [...] It remains indivisible, not because it cannot be divided but because the question of divisibility is no longer posed concerning it" (Minkowski, 1970, p. 35). Being "a now that is stretched out" (ibid., 36), it is not point-like but, thanks to a duration[6] whose limits are very fluid, it coincides *de facto* with existence. Indeed, as we already know (cf. especially Chapter 6), Schmitz elects the primitive presence/present[7] as the principle of his new phenomenology (Schmitz, 1964, p. 149), using it as a real "mythogenous idea" or, to use a less critical expression, as "a mere metacategory we could also call awareness" (cf. Stepath, 2006, p. 119, p. 123).

I simply want to underline here that for Schmitz "to be means to be-present" (Timm, 1993, p. 262), because the person is a conscious subject with the capability of self-ascription if, and only if, they are always able to go back to their proto-identitary life. This original life, based on "subjective facts" which are subjective even "in a more original way than the subjects" (Schmitz, 1996, p. 26), and are therefore identical (vertically) to us without reflective identification (horizontally), is exactly that of the primitive presence/present, to be understood in both a temporal and an (older) spatial sense (Schmitz, 1964, p. 196), a bit like the Japanese spatial-temporal *ma*. According to a felt-bodily economy based on the vital drive and on affective involvement, fright (exemplarily), pain, anguish, shame, orgasm and panic[8] generate a sort of contradiction – the self would like to escape itself! – that leads to catastrophic reactions (also in Plessner's sense). This impossible escape, unlike the case of deep joy (in which one merges with the self's other) or the simple prereflective going on existing (Andermann, 2007, p. 273), shows precisely through a felt contraction that we are spatio-temporal attempting to escape from the spatio-temporal present/presence (the here and now). Which, indeed, proves that we *are* here and now: what happens is about us, it concerns us.

As I have already considered this theory, here I will only indicate some key points. First of all, this primitive present is the fusion point of five elements (here, now, being, this and I), which however are recognized as five only *ex post*, namely by an unfolded subjectivity (Blume, 2003, p. 42, fn. 4). The certainty provided by the primitive present, guaranteed by a felt-bodily resonance, implies no certainty about our (self-attributed and somewhat abstract) properties and the real nature of what appears to us, but only about our being emotionally concerned as subjects. It serves therefore as the deepest *principium individuationis*, namely as a felt-bodily individuation without singularisation, which is always a subsumption under categories. The immediate evidence of reality, given through a sort of tightness of the felt body, obviously applies also to animals and newborns, to anyone who can "take fright" or, in other words, experience the breakdown of the continuum due to something sudden and new without still resorting to singularisation or self-ascription. On the contrary, the self-ascription through which my identity is normally explained is actually only possible (to avoid a *regressus ad infinitum*) through a self-consciousness without identification, that is, only if I am already acquainted with myself through this (primitive) present.

Together, the five points, with the resulting affective involvement in the form of a felt-bodily breakdown and contractedness, are "absolute" because they are not yet related to their five respective (but only relative, as we'll see) elements. This "presingular subjectivity or self-awareness", which is the undeniable guarantee of the coincidence between identity and subjectivity, can and must have a development,[9] even if every later situation of the unfolded present is nothing but a particular sphere "guaranteed" by the primitive present and traced around the "vital phenomena affecting my whole being" (Minkowski, 1970, p. 125). The unfolded present (i.e. the world), which in a way can never be fully achieved, explicates meanings by singularisation, frees them – also by means of sentential speech – from their original complete subjectivity and thus enables the self to distance itself from the original absoluteness. Indeed, it transcends the felt-bodily "here" (hence polarities like absolute location vs. wideness and space as a system of relative location) and "now" (hence absolute moment vs. duration as a modal temporality and the distinction between not-yet-being and no-longer-being).[10] It goes beyond the "existence or reality" by projecting itself into the mere possibility and leaves behind the "this" by relating to identity and difference and even looking ahead to the future. Lastly, it can distinguish between one's own world and another's or an alien world. In a word: through the fivefold unfolded present, the human being (unlike the non-human animal) doubtlessly goes beyond the present situation. But the most important thing is that only through the collapse of personal emancipation,[11] caused by fright but also laughing and crying, and through the resulting regression to the primitive present (personal regression), does the subject feel and know with certainty that it exists. Only when meanings suddenly fall back into their internally diffuse meaningfulness and we are affectively involved, do we have full confidence in reality.[12]

But I have already abundantly dwelt on this idea of the oscillation between regression and personal emancipation. To move on to another key point, Schmitz differentiates time into three less original modalities, that is, pure modal temporality (being, not-yet-being and no-longer-being), pure relational temporality (relation of earlier to later or simultaneity) and finally the blended mode of modal relational temporality, which is the basis of the flow of time (the past constantly grows, the future constantly shrinks and the present is constantly shifting) but also of its well-known aporias. In the first case (pure modal temporality) the present is different from the past but so exposed to the ingression of the new (future) as to be replaced by appresence (*Appräsenz* nel senso di *ad presens*) as the fusion of present and future. The qualitative bipolar structure (past and appresence) of the primitive present (the now) is then quantitatively levelled in the unfolded present in the form both of modal relational temporality (past present, present present, future present) and of pure relational temporality (time of the day and clock time). However, it would be a mistake to absolutize one of these temporal structures, thereby sacrificing the paradoxical experience of everyday life that is neither wholly unconscious (as Stepath, 2006, p. 132 seems to think) nor a return to nature (Soentgen, 1998, p. 52). The present in fact usually manifests itself both as the shocking absolute now (modal temporality) and as the relative "nows" in the succession of events (relational temporality), but especially as the lifewordly fusion of the previous two modes (modal relational temporality) that characterises our *experientia vaga* – whose best analysis, at least since Baumgarten, is aesthetics.

The next step is to evaluate the link between presence/present and felt-bodily dimension. Indeed, it should be recognised that Gumbrecht's presence and Schmitz's present converge towards a full phenomenological rehabilitation of all that appears as such and we directly felt-bodily perceive. Indeed, the undoubted differences between the two – whereas the former looks to a lasting situational and somewhat more abstract presence, the latter is more concerned with a momentary, unrepeatable and somewhat more shocking present – must not be overstated. They have a shared interest in tightly linking together perception and imposition of presence (cf. Wiesing, 2014, pp. 79, 85, with the exception, of course, of what is temporally but not spatially present to us). Presence is not always perceived in a strictly sensorial way: sometimes (think of a speaker on the podium or an actor on stage, for example) we certainly felt-bodily perceive the atmospheric presence (and thus the feedback) of our outside, of the world, as if it were a gaze towards us and our presence were, as a consequence, our being looked at by them (Hauskeller, 1995, pp. 115–118). This brings us to felt-bodily presence, which today seems to be socially repressed, to put it simply, because of the ubiquitous virtualization and the increasing obsession with (also bodily) privacy. However, on closer inspection, this is only partially true. Indeed, it is true that a video conference cannot fully replace a face to face meeting, that only through bodily presence one can fully (synaesthetically) assess the "how" of another's talking, or that only a journey that does not rely entirely on the goal gives a chance to experience and appreciate a place (its presence). However, it's also true that technical development might unleash unexpected felt-bodily potentialities for the very reason that they are not bodily in a strictly physical sense.

This approach cannot be discussed in detail here, but one can see that it is rooted in a neo-phenomenological theory of felt-bodily communication. In fact, we felt-bodily communicate with everything that is other (animated or not, cf. Schmitz, 2017, pp. 64–79 and Griffero, 2017c) because we feel its presence-present (its "voluminosity") through our felt-bodily presence, that is, through a resonance understood as one of many possibilities contained in the intracorporeal economy of contraction (encorporation) and expansion (excorporation). By virtue of this simultaneous presence of the communication partners, and regardless of whether the subject thus embodies or is disembodied into something, it can be argued that feeling the presence-present does not indicate anything mystical or too performative. In other words, a real felt-bodily (in a way, extended) presence-present of ourselves and, at the same time, of the outside world is guaranteed not only by extreme and almost always solipsistic experiences or by some sophisticated training (meditation etc., cf. Böhme, 2010, p. 127; 2014, pp. 152, 154), but also by our ordinary everyday experience (walking down a street, contemplating a landscape, waiting for the train, even feeling our heartbeat), provided that one is able, in the short term at least, to carelessly open oneself to the moment, get out of the usual goal-oriented life and feel oneself in the (even) ephemeral *qualia* of the world.

As can be seen, I have doubts about every too-performative and voluntary approach to the felt-bodily presence-present.[13] To conclude, I must underline once again that being-present-and-in-a-presence always means being-in-a-mood, since "a mood contributes to sensing where we are. By feeling our own presence, we feel the space in which we are present" (Böhme, 2017a, p. 137). In other words, we cannot pay attention to the presence-present without also paying attention to the presence-present of our existence.[14] Thanks to this, we feel a relatively stable atmosphere poured out in the (lived) present space and time, which gives a quasi-thingly and quasi-objective tone to all other partial and ephemeral moods and makes them something undoubtedly real and actual (effective *qua* affective in Böhme's sense). Against the highly predictable argument that this (relatively extended) presence-present could end up in solipsism,[15] one ought to underline with Schmitz that, being the "person so open that one can no longer speak of an enclosed private internal world" (Schmitz, 2008b, p. 31; cf. also Schmitz, 2017, pp. 148–161), this expanding present-awareness (in every sense of the term)[16] is something aesthetic *and therefore* ecstatic precisely because it always "learns" from what is other but present in the present world.

But is this felt-bodily presentism only the "pie in the sky" of not-so-young people who wish to remain forever youthful and escape the hyper-technological present? I undoubtedly agree with Gumbrecht's claim:

> Don't those of us who are older have the right to remain on our islands for as long as possible? Why should we clumsily adapt to the demands of the electronic that dominate the new present? We are already living in a vast moment of simultaneities. There is no need to reject us – we who embody one of many pasts – from our havens in the broad present.
>
> (Gumbrecht, 2014, p. xiv)

Any serious reflection on presence-present cannot help putting an end to today's obligation to be constantly available, too often misunderstood as infinite (pseudo-democratic and ultimately enslaving),[17] and thus keep alive the principle according to which presence emerges and grows only thanks to its existential contrast with absence. But, at this point, neurophysiologists inevitably step forward and, brushing off every phenomenological and cultural-critical issue, remind us with a condescending look and thin considerations – but let's remember that the young Husserl sharpened his penknife until it broke! – that what we call outside presence is nothing but a construction (image) of the brain. Yet the brain itself, which is no doubt part of the outside world, would therefore only be a construction (image) of the brain, and this would be nothing but an image of an image, and so on. It is always the same old story … but now even without the good old transcendent (Platonic) consolations.

Notes

1 But "it is better to suffer from an unfulfilled desire than to lose desire altogether" (Gumbrecht, 2014, p. 9).
2 Rituals are "attempts to correspond to cosmological frames" (Gumbrecht 2014, p. 3).
3 "Aesthetic intuition is a radical form of residency in the here and now" (Seel, 2005, p. 33).

4 "'Reading for *Stimmung*' always means paying attention to the textual dimension of the forms that envelop us and our bodies as a physical reality – something that can catalyze inner feelings without matters of representation necessarily being involved" (Gumbrecht, 2012b, p. 5).

5 According to this deictic attitude, "words are experienced as pointing to things rather than standing 'for them'" (Gumbrecht, 2014, p. 6).

6 We know (better, we feel) this duration without necessarily accepting the neuroscientific proposal to fixate it on a brain interval of three seconds.

7 He thus assumes and expands some concepts proposed by Eugène Minkowski. The "I exist" – "a time of halting in our activity" that is neither purely dynamic nor purely static but "an absolute point in space, precisely the one where I actually am and which is a true center of the world for the active ego" (Minkowski, 1970, pp. 124–125) – is translated by Minkowski into the absolute spatio-temporal point me-here-now (*moi ici maintenant*). But there are also analogies with Volkelt's absoluteness of here-now-thisness (2013, p. 50 ff.). Cf. Stepath (2006, p. 123, fn. 475).

8 The examples provided by Soentgen (1998, p. 50) – dismissal, fascination, etc. – are more controversial, inasmuch as they refer to a less original felt-bodily state.

9 Also in an ontogenetic sense: for an application of this to child learning cf. Schultheis (1998, p. 93 ff.).

10 For an analysis of time based on the critique of the unilateral development of the primitive present/presence in a relational temporality (instead of a more appropriate modal relational temporality) see Schmitz (1990, pp. 247–274; 2019, pp. 83–86.

11 "People […] do not simply present themselves as if they were new guests to a world already definitely divided into wholly individual things, but the very fact that they present themselves is an aspect of a fivefold structure of the primitive present/presence in a world endowed with its own form" (Schmitz, 2003, p. 18).

12 "Reality appears all of a sudden, breaking the duration of subsisting in primitive present/presence (for example, when a sudden and intense noise wakes one up from dozing)" (Schmitz, 2005b, p. 29). I have already spoken about "good" regressions in this book (cf. Chapter 6).

13 This makes Shusterman's somaesthetics (see at least Shusterman, 1999, 2008, 2012) an important and productive playing field (Griffero, 2017b, pp. 160–161) that certainly deserves to be debated thoroughly.

14 The intuitive awareness of presence implies "an aareness of *a* here and now that also encompasses an awareness of *my* here and now" (Seel, 2005, p. 32, emphasis added).

15 See for example Irrgang (2016, p. 100) and Kersting (1992, p. 83), for whom "the presentist is the solipsist's brother in the context of a philosophy of time". I believe it is also a bit misleading to define the presence-awareness an endogenous outcome (as Böhme, 2014, p. 157 does).

16 "The present is the sum of representations and (physical as well as spiritual) sensations we consciously and unconsciously alternatively perceive, and at the same time always the horizon of our perception. Our access to the world is rooted in this process of discontinuous oscillation. If this access is optimistic, one links the present to the representation of a 'kairocracy of life'; if it's pessimistic, one understands the present as 'death drive'" (Schmitz, 2002, p. 27).

17 "How impossibly old-fashioned is it if I regularly feel that in this type and under these conditions of interaction it should be exclusively my privilege to be 'available' or not?" (Gumbrecht, 2014, p. 63).

Part Three
Atmospherological exercises

Two premises

In order to explore some atmospheric fields, I can no longer postpone the clarification of some preliminary points, relating both to the distinction between different types of atmospheres, and to the variety of possible interactions between the perceiver and the atmospheric sentiment, often referred to in the previous pages. I must therefore make two important points.

1) In a nutshell, my ontologically inflationary conception consists in differentiating between three "idealtypes" of atmospheres. On the basis of their decreasing objectivity (as well as intensity) and increasing objectual focalization,[1] atmospheres can be, first of all, *prototypical*, i.e. objective, external, unintentional, sometimes lacking a precise name, relating to the lived space, and therefore broadly equivalent to the *Stimmungen* of the German philosophical tradition. But there are also *derivative-relational* atmospheres, which are objective, external, intentionally produced, related to directional space and implying an "in-between" between subject and object. And there are even quite *spurious* atmospheres, which are subjective, projective and even related to single objects and local spaces. This wide range of atmospheric feelings has therefore, as its extreme poles, groundless quasi-objective atmospheres and merely projective individual moods.

I have used the term idealtypes. For me, this means that this three-way differentiation of atmospheres must always be considered flexible. In fact, by changing perspective and felt-bodily encorporation, a prototypical atmosphere may also become a derivative-relational one, such as when, for example, an atmosphere breaks free of its anchoring point (which is the source of its radiation), loses its thematic reference and materializes into random or at least inappropriate things and situations. Such an atmosphere may even become a simple emotion thanks to a precise point of crystallization. Likewise, an idiosyncratic mood may increasingly (albeit deceptively) dismiss its original attitude, or generate a specific interplay with this or that object transforming an initially spurious atmosphere into a derivative-relational one, and even objectify itself into an atmosphere experienced (almost) as powerful as the prototypical one. And so forth, with other fluid transitions from one state to the next. In fact, a subjective mood may also be already generated by a previous (pre-formatted in a sense) atmosphere, which in turn might have been triggered even by repeated emotions of the "same" type – such as when a frustrated anger, for example, generates a mood of *ressentiment*, which can consolidate into a personality trait, creating a resentful person.

Similarly, if an atmosphere subjectifies itself into a mood, for example when entering one's personal involvement,[2] it is likely that one could not perceive that atmosphere without a moody (pre-formatted) background. Well, this alternation between three kinds of atmospheres only exemplifies the more general intertwining of different affects (emotions, moods, atmospheres, etc.). This is the fluid process of focusing-foregrounding and blurring-backgrounding[3] according to which, to stick to an elementary example, a mood could (spatially) become more focused and thus turn into an atmosphere, whereas an atmosphere could in turn blur out of focus and thus turn into a mood. What is atmospherologically important, though, is not to conceive of this fluid temporal process of backgrounding and foregrounding as something entirely depending on the subject's arbitrary affective state.[4]

2) Also, it should be remembered that the relationship between the perceiver and the three types of atmospheres may result in many different emotional interactive "games". Roughly put, (a) an atmosphere can overwhelm a perceiver (ingressive-antagonistic encounter), completely reorient their emotional situation and be refractory, as a real antagonist, to a more or less conscious attempt at a projective, reflective and amending interpretation. (b) It can find a perceiver to be attuned with it (syntonic encounter), to the point that, whereas others may even be upset by the atmosphere in question, they do not even realize they have entered it. This apparent absence of atmosphere can be explained by a (Gestaltic) insufficient contrast between figure and background, thus causing an embarrassing affective and social inadequacy in the unaware perceiver, but also by a felt-bodily disposition so compact as to prevent them from abandoning themselves to the situation. The most powerful atmospheres, however, are exactly those which lie in the background and provide the tacit and non-thetical contextual conditions for transitive perceptions. (c) Sometimes one clearly recognizes an atmosphere in its "objective" roots (mere contemplation), but without being really felt-bodily involved. This emotionless recognition, while being further proof of the failure of the projectivist thesis, also raises the controversial question (that cannot be addressed here) of how exactly emotion and cognition overlap or interact.

(d) An atmosphere can also elicit a resistance (mood protest) and the very intensity of this emotional reaction is indeed the best proof of the objective effectiveness of the atmosphere one reacts to. (e) An atmosphere, which as such is supervenient on its (material and immaterial) generators, may sometimes be aroused by elements normally expressing completely different moods, such as when our self-disdain is strengthened by the (however suggestive) beauty of a landscape. (f) Although the prototypical atmosphere is that suggested by the first impression (Griffero, 2014a, pp. 29–31), the subsequent ones may give rise to relatively different atmospherizations (though influenced by the very first one). There are various reasons for this: perceptual and climatic changes, end of sensory illusions, change in the perceiver's physiological conditions, contrast between the atmosphere and its unwanted sub-atmospheres, additional and divergent cognitions, purely idiosyncratic experiences, focusing over time on misleading spheres of condensation (thus also infecting objects that are not responsible for that feeling, such as when the fear of the dentist spreads a negative atmosphere on every visible object of their waiting room, etc.). Even if a non-prototypical atmosphere can therefore change over time and be filtered through the perceiver's new evaluations, this does not mean that one should embrace projectivistic relativism. And *pour cause*: if people always perceived things subjectively, no atmosphere different from (let alone contrary to) the perceiver's mood could ever overwhelm and affect them. (g) An atmosphere, finally, might elicit an active feedback loop to the extent that perceivers, for example the spectators of a performance coordinate themselves with it so as to reinforce the staged atmosphere.

Considering these two theoretical but also heuristic-methodological premises, we can now venture out into the open sea, trying to apply the spirit of this pathic and atmospherological aesthetics to some concrete areas, such as architectural design and dwelling, the spirit of places and natural expressiveness (of the earth, in this case), the democratic climate, the pedagogical environment and the staging of atmospheres (in the theatre). The first atmospherological "exercise" concerns architecture, which has always been, non-coincidentally, a privileged field to understand what emotional spaces may be.

Notes

1 Further criteria may be greater or lesser intensity, awareness, etc.
2 Tellenbach (1968) claims something similar by distinguishing between atmosphere as impersonal reality and *Stimmung* as unity of I-feeling and world.
3 Inspired by Musil's (1995, II, p. 1307) suggestive idea: "there is no 'mood' that does not include specific emotions that form and dissolve again; and there is no specific emotion that, at least where it can be said to 'radiate', 'seize', 'operate out of itself', 'extend itself' or operate on the world 'directly', without an external emotion, does not allow the characteristics of the nonspecific emotion to peer through".
4 See Bulka's (2015, p. 319) right objection to Goldie (2000, p. 148).

9 Architectural affordances

The atmospheric authority of spaces

"Each thing says what it is and what [we] ought to do with it: a fruit says, 'Eat me'; water says, 'Drink me'; thunder says, 'Fear me', and woman says, 'Love me'" – writes Koffka (1999, p. 7), with a fictional primitive man in mind. Regardless of the sexist remark (those were different times, after all!), one should note how Koffka highlights here the so-called demand character, or invitation character and valence of our environment – a character that doesn't completely change according to the need or the intention of the agent. But couldn't this be applied more generally, and *a fortiori* especially in an architectural environment, conceived not as a collection of causes but as an emotional set of possibilities of action?

The point is that, like every physical *Gestalt*, buildings, according to Wölfflin, "have a character only because we ourselves have a body" (Wölfflin, 1994, p. 151): that is, they communicate their qualities to us because they are the same qualities with which we express ourselves (weight, hardness, etc.). It follows that life's great feelings and atmospheres are based on the premise of the lasting condition of our bodily organization (ibid., pp. 157–158). This also explains the organic-kinaesthetic effect of visuality, and especially of symmetry and asymmetry on breathing and muscles.

> powerful columns produce energetic stimulations in us, our respiration harmonizes with the expansive or narrow nature of space. In the former case we are stimulated as if we ourselves were the supporting columns; in the latter case we breathe deeply and feel as if our chest were as wide as the hall [...] The architectural impression [...] is essentially based on a directly bodily feeling.
>
> (ibid., pp. 154–155)

Wölfflin's late 19th-century approach is of course theoretically weak because of its trivial impression/expression dualism, which unfortunately is partly still projectivistic and at times very naïf (like Freud's), as it recurs in associations mostly based on anthropomorphic images only (weight-nostalgia, symmetry-wellbeing, windows-eyes, round arch-happiness, etc.). However, it correctly traces formal perception back to how one affectively and bodily feels oneself in the presence of something else (Böhme, 2006, p. 110). Through these mimetic-kinaesthetic and semi-automatic reactions to (also atmospheric) situations, which like a sort of "cognitive unconscious" are the more perfect the less one thinks about them, the felt-body senses tensions and expressive orientations in its surroundings, and especially in a lived space as special as the architectural one, whose configurations we indeed unconsciously mimic with our bones and muscles (Pallasmaa, 2005, p. 67).

Many researchers have recently emphasized that objects and spaces are not merely functional but always affect the perceiver-agent emotionally, making certain behaviours more likely to occur. First of all think, as mentioned (cf. 2.3), one can think of the internal reproduction of movements,[1] of directional tensions immediately and physiognomically felt in lines and forms (Klages, 1964, p. 447 and Koffka, 1999, pp. 359–364), even static ones, in which "the felt body, which we feel immediately, and the perceived form perfectly coincide" (Schmitz, 1965, p. 42). This is the well-known non-projectivistic *Gestalt* principle of isomorphism (cf. Arnheim, 1966, p. 64), due not to "anthropomorphic metaphors of action, but [to] an adequate reproduction of the motor suggestions that come from the forms and are immanent in them" (Schmitz, 1965, p. 38). In short: we find something about and in things which demands our acceptance and we no longer transfer it into the things, but in some way we apprehend it from the things themselves (cf. Heidegger, 1995, p. 85). Atmospheres then seem to be bridge-qualities, founded on a felt-bodily communication without any real contact between user and object, and their suggestions are perceivable as virtual movements. No "synthesizing" subject could give rise to qualities where there are none: no one can give melancholy to a building if this melancholy does not already belong to the building itself.

But what is of utmost importance, especially in architecture, no longer conceived only as a visual art but as an art of invitation and promise, is that atmospheric perception is direct and deambulatory, kinaesthetic and affectively engaging, synaesthetic or at least polymodal:[2] a kind of "peripheral unfocused vision" (Pallasmaa, 2005, p. 10; cf. 2014) that often we facilitate by semi-closing our eyes. Its outcome is a holistic and emotional being-in-the-world, or even a deeper contact, if it is true that all the senses are but an extension of hapticity. So, one could even say that atmospheric perception is more being-perceived than a perceiving: a primordial, anonymous and ante-predicative pact with the world. A pact that, especially in regard to architecture, is not confined to the "distancing and detaching" (Pallasmaa, 2005, p. 22) sense of sight, and whose result is precisely the atmospheric or, if you prefer, even a "landscape" in Straus' sense of an affective and non-gnostic way of being-in-the-world (cf. Straus, 1963, p. 322 and Hahn, 2012a, p. 18). In this sense, to perceive an atmosphere always means to co-perceive our affective felt-bodily situation and to ascertain how we feel in a certain place.

Here I am clearly upholding the anti-separatist thesis on the relation between perception and value[3] as well as the idea that, by feeling,[4] the perceiver, who in a way is both the subject and object of perception, perceives atmospheric affordances. The latter do not only imply a preconscious reflective-motor response or, if you want, a "handiness" (Heidegger), but "convey messages about their possible uses, actions and functions" (Norman, 2002, p. 82). What I mean to say with this ecological account of atmospheric agency is that humans are not surrounded by things that are devoid of meaning, but by things and quasi-things that are always already affectively connoted. As James Gibson put it, "the perceiving of an affordance is not a process of perceiving a value-free physical object to which meaning is somehow added in a way that no one has been able to agree upon; it is a process of perceiving a value-rich ecological object" (Gibson, 1986, p. 140). In the human lifeworld there is indeed nothing rarer than the perception of an inexpressive object, and it is perhaps atmosphericness itself that makes it possible for mere sensation to become real perception. Going beyond Gibson, it must be said that forms, whether static or in motion, do not merely express apparent causal relations and pragmatic affordances but also tertiary or sentimental (and therefore atmospheric) qualities that permeate the space in which they are perceived. They are, in other words, "perceptive ingredients inside facts themselves", "shivers of meaning present in things [or quasi-things]" (Bozzi, 1998, p. 16 f.), no less inter-observable and repeatable than perspective properties (within certain limits).

These pathic affordances, responsible for our spontaneous-intuitive evaluations, are aroused by accumulation centres filling – in both a causal-topologic and an axiologic and motivational sense (salience, preferability, etc.) – the multidimensional sensory continuum that makes up the world of existence. It is in this (also atmospheric) sense that, following a saying usually attributed to Wertheimer, "black is lugubrious even before being black". The idea – very briefly – is that atmospheres function as (amodal) affordances, i.e. as ecological invitations or meanings that are ontologically rooted in things and quasi-things, namely as demands that are not only pragmatic-behavioural and visual. While the environment can invite a certain action or even urge a person to do something, to an atmospheric affordance indeed one does not necessarily react with a given behaviour. It's certainly true that performed actions, providing that they are not regarded as computable "functions" but as sensual and communicative embodied forms of use, always have a pathic meaning,[5] and that constantly different uses contribute significantly to the production of a specific atmosphere. But it is equally true that one may also react to atmospheric *qualia* with an (also aesthetic) distance, in the sense that we can feel the atmosphere of a chair, for example, both by sitting on it and by perceiving it from a distance.

Moreover, perceiving an atmospheric affordance seems to "demand" a special objectivity, just like the Kantian judgment of taste: for instance, it is legitimate to expect that the unease and the feeling of being spied on aroused in us by a dark wood (a very different sensation from Kant's disinterested and de-sensualized judgment) is affectively and corporeally felt by anyone who shares such an experience. The disquietude can certainly be greater for a city person than for a lumberjack, but certainly the wood has, for everyone, traits that impede a projection of joy or light-heartedness. In opposition to the associationist temptation or, even worse, the conventionalist one, I must restate that the atmospheric affordance of disquietude aroused by the wood does not derive from the thought of fear, but is rather the immediate irradiation of a quasi-thingly feeling that is spatially poured out. Association, if anything, comes after and is certainly not arbitrary, given that "phenomenically, it is not the reference that produces the atmosphere, but it is the atmosphere that makes the reference possible" (Hauskeller, 1995, p. 139).

While steering away from an (overly postmodernist) apology of the so-called patchwork-cities and of the "prostitution of atmospheres" arising from them (cf. Meisenheimer, 2006, p. 136), I nevertheless must stress that design, as the *mise-en-scène* of felt-bodily affordances, should not disregard "how we feel" in the designed space. For example, it cannot underestimate that even the shadow cast by a building, invading the neighbouring space, determines a powerful aesthetic and felt-bodily (and therefore atmospheric) effect. Similarly, providing that smells are signs of identification within the city, it is an architect's job to make sure that certain odours contribute to the urban atmosphere they mean to suggest (cf. Bischoff, 2005). This applies even more to interior design. For instance, in a shop there can be a warm atmosphere in which rusticity and genuineness invite us to linger, to look around (local products, wooden shelves, aesthetizing disposition of the goods, suffused lights, kindness of staff, minimization of strictly economic elements like prices and counters, etc.). Or there can be a cold atmosphere, in which anonymity and serial production suggest, instead, to walk through it rapidly and with the sole purpose of purchase (multinational brands, metal or Plexiglas shelves, disposition of the goods by price and target, central and direct lighting, detachment and impersonality of staff, visual highlighting of prices and counters, etc.).[6] Why do we almost always prefer a wooden object to a plastic or metal one if not for its warm atmosphere? If not for its ability to evoke, for example, life (the tree with its roots), peace (the slow and regular life of the tree), strength (under large trees we feel protected), sensuality (suggested both by touch and by smell), industriousness (which urges us to do something), charm (also hypnotic, of its veins, for example) and the past (almost as if the wooden object preserved the memory of past environments)?

In short, the charm of materials – or of their "consonance", conceived as the right distance from each other (cf. Zumthor, 2006, pp. 23, 27) – lies largely in their affordances, that is, in their atmospheric potentiality. Especially when they are connected to "old, half-forgotten memories", and therefore contain what one could define "the deepest architectural experience" (Zumthor, 1998, p. 10), materials create a real *genius loci*, namely an experience that today is largely overshadowed because all too often "the 'pure' place turns into something non-spiritual" and "'pure' spirit into something that has no place" (Kozljanič, 2004, II, p. 264). Just like the atmosphere, the *genius loci* is not "a secondary moment or maybe even a parasitic character" of physical data, nor is it only the epiphenomenon of subjective associationism. Vice versa, it consists in making a space into a place, namely in discovering the potential sense present in a given environment (cf. Kuhlmann, 1998), precisely in the form of an overall atmosphere. When space has or acquires an atmospheric charge, it expresses a *genius loci*, regardless of whether it concerns archetypes of natural places or whether it relates, at least in part, to various cultural-geographical perspectives and affordances (as we will see better later: cf. 3.4).

Now let's move on to the atmospheric affordances of architecture.[7] Truth be told, there is no need to say much about the general relationship between architecture and atmosphere, because the purpose of architecture as *Gesamtkunstwerk* is largely seen as that of framing and structuring our felt-bodily dwelling and, generally, our affectivity in a place. Suffice to say that "architecture is but a stage set that produces a sensuous atmosphere", or even that it "is defined by atmosphere" (Wigley, 1998, p. 21, p. 27), because "the production of architecture (of the 'architectural space') is the production of feelings" (Hahn, 2012b, p. 79). Also: architectural atmospheres are responsible for "immediate understanding, immediate contact, immediate rejection", and are generated fundamentally by everything: "things, people, air, noises, tone, colours, material presences, structures, also forms" (Zumthor, 2006, p. 13, p. 17). They may even arise through sound or echo, because "every building or space has its characteristic sound of intimacy or monumentality, invitation or rejection, hospitality or hostility" (Pallasmaa, 2005, p. 50)[8]: "there are buildings that resonate wonderfully, that tell me: I am safe, I am not alone" (Zumthor, 2006, p. 33).[9] For instance, think about the bureaucratic coldness hinted at by a wholly interiorised building (made of glass or granite). In short, by generating "orientations, kinetic suggestions, markings" (Böhme, 2006, p. 113), buildings produce a wide range of atmospheres and, in their capacity as authentic scripted spaces, they force the perceiver to immerse themselves in them (cf. Bieger, 2011). Architectural atmospheres pathically modulate the pericorporeal space of the observer, and they do so in a consistent manner, because unlike other more transient qualities, the urban and architectural forms stably give rise to certain atmospheres.

Embracing the transition from a semiotic paradigm to a performative one (cf. Jäkel, 2013, p. 32), I ignore *a limine* the trivial and insufficient communication model (sender-message-receiver) and replace the model of sign interpretation (as if the world were a text) with the model of embodiment (or felt-bodily communication) as the result of an exchange process between humans and things. This is the model of co-presence, based on the (in a sense) porosity of the buildings and on the fact that when I place myself in a space, the space settles in me. But I also reject both Western ocularcentrism and the idea that architecture is just the creation of large sculptures (cf. Böhme, 2006, p. 175). Instead of comparing the architectural experience to the mere (fixated and focused) observation of a sequence of retinal images required by its "frontal ontology" and often necessarily witnessed only as photographic image, I prefer to start from a deeper and pathic idea of a "utilitarian" sense of architecture, which is not primarily "either a message or a symbol, but [...] an envelope and background for life which goes on in and around it" (Zumthor, 1998, p. 13). The architectural atmosphere, even if understood as the "effect" (Camillo Sitte) or "imageability" of a city (Kevin Lynch), is therefore something that is not seen but *felt* and co-produced: "feeling our own presence is feeling the space in which we are present" (Böhme, 2006, p. 110), because to feel means both to be situated in a space and to feel in a certain way. And this way of feeling certain space affordances permeates every other following experience, even in a deep and hypodermic way.

One can put it in many different ways. One could simply say that "forms are not bound by their physical limits [but] emanate and model space" (Giedion, 2002, p. xlvii), that they extend as ecstasies in a certain space, thus articulating the sphere of their presence and taking away "the homogeneity of the surrounding space and [filling] it with tensions and suggestions of movement" (Böhme, 2017a, p. 19). Also, the special things we call buildings "always convey an intense feeling of their spatial quality. They embrace the mysterious void called space in a special way and make it vibrate" (Zumthor, 1998, p. 21) and are therefore inextricably linked, as we shall see later (cf. 3.2), with domosphere: indeed "dwelling is feeling to feel oneself, perceiving to perceive oneself" (Vitta, 2008, p. 97), feeling at home, namely an outstanding example of encounter between *aisthesis* and *pathos*.[10] But how does the beholder really meet an architectural atmosphere?

One of the best ways to explain it is to conceive this kind of atmosphere as the spontaneous and circular intertwinement between the repertoire of architectural gestures and the user's repertoire of felt-bodily and spatial experiences: the former relies on the expectations of felt-bodily feelings[11] and the latter depends on the specific gestures of built-up areas, in the sense that felt-bodily gestures are here a direct response to architectural forms, perceived as forms that give us a lived "direction". But "gesture" is the important word here. Indeed, in architectural forms, which are objects that oscillate between being a thing and an event,[12] we can read some primary gestures, which are not only inspired by organic shapes and whose pathological reverse, however, lies in the main spatial phobias. These gestures are not properly things or elements of the buildings, nor are they precisely circumscribed *Gestalten*: without being clearly visible entities like walls, windows and doors, they are intrinsic forms, or, better, less static and cultural codified "figures" (Jäkel, 2013, pp. 110–112) of the "surrounding felt-bodily reality" (cf. also Dürckheim, 2005). Through their force fields they suggest (and imply) some motor figures of the user and thus give life to a real movement and to an atmospheric dramaturgy. A window, for example – and not every window, but this particular window because of its special way of directing the gaze outwards (Pallasmaa in Borch, 2014, p. 99) – affects the felt body through its viewing direction, its soffit, its materiality, etc, in short through some acts. That's why "fundamental architectural experiences are verbs rather than nouns" (Pallasmaa, 2011, p. 53).

Let's take, for example, the primary gestures identified by Wolfgang Meisenheimer (2006, p. 25 ff.). They are: a) the erection, i.e. the vertical attitude with all its felt-bodily and symbolic-pathic effects (resisting the passage of time, overcoming gravity and death, demonstration of masculine power, etc.); then b) the distinction between here and there as absolute places, which always produce very different atmospheric "isles"; then c) marking the borders between inside and outside through all the possible frames (walls, frames, threshold, etc.) in order to create real inner landscapes; and lastly d) various ways of developing the fundamental fluctuation between contraction and expansion (synaesthetically produced). These primary gestures are relatively invariant and devoid of any arbitrariness. Think, for example, of distancing (from the noise, a bad smell, danger, a harsh climate, etc.) as the main lasting effect of architecture on the felt body (ibid., p. 62 ff.). But think also, in more in detail, of the acute angle of a building, which always radiates a sheltering mood within but a hostile one outside, also because it isolates the building from the others and, appearing as a sign of power, excludes passers-by (cf. Hasse, 2012, pp. 101–112).

Since "spatial gestures suggest certain movements, behaviours, actions and even moods to us" (Jäkel, 2013, p. 15), the architectural ones, conceived in their various functions,[13] are therefore a good way to understand how we perceive the architectural space. The architect is able to convey these architectural gestures (no matter if thanks to optical and even tactile illusions)[14] by means of a kind of scenic thought. This way of thinking conceives the architectural space as a choreographic score and the atmospheric perception as a specific movement (think, for example, of the fearful approach to St. Peter by Bernini's colonnade). This way the architect relies on the perceiver-user as a partner of this spatial directing. A wall, for example, accompanies me for a while, then guides me to a corner, rises high above my eyes, I follow it and both its process and its nature determine the way I walk. In this sense the wall has really physiognomic or expressive properties, which reside in a sense "within" the object as its own most "essential" meanings (Metzger). But a door is no less atmospheric-gestural, even if it does not lead anywhere and for this reason forces us to an engaging kinaesthetic epoché (cf. Waldenfels, 1996). Nor should we forget that the constructed space sometimes moves the beholder not thanks to precise signals but thanks to a certain ambiguity and certainly thanks to a bilateral vagueness: namely both due to a real loss of focus by the subject and to a kind of secret and mystery inside the object itself, in other words thanks also to a modulated and gentle immersivity.[15]

Leaving aside a more detailed analysis of how to exhaustively describe the architectural gestures[16] and forgetting for a moment the kind of atmosphere particularly generated by gestures,[17] let's ask this question: if architectural gestures invite us to explore the spatial continuum and to expose ourselves to its atmospheric affordances, what are the basic types of this atmospheric encounter? Here I am referring freely to the aforementioned atmospheric "games". Well, even in the case of architectural experience one can experience an ingressive encounter with a prototypical atmosphere refractory to any attempt at a projective re-interpretation or at reflective reductionism: this encounter consists in entering a building and passing from one atmosphere to another, usually from a physical atmosphere to an artificial one.[18] A prototypical atmosphere consists therefore of a spatial-affective presence that is clearly distinct from the perceiving subject and unilaterally affects them. It is something that "is simply there [and] on which we cannot focus our consciousness" (Zumthor, 1998, p. 17): something that can even be compared to waking up (cf. Hahn, 2012b, p. 87). Or else, one might experience an architectural atmosphere as syntonic thanks to the coincidence between the form of perception (the mood with which we perceive) and its content (the spatial mood perceived). Here one could speak of immersion, provided that the term (cf. also Pinzer, 2013) means a felt-bodily (not necessarily artificial) involvement, thanks to which whoever is grasped by the atmosphere then perceives themselves and their everyday relationships less vividly than before.

A third possibility is that the architectural atmosphere may instead be antagonistic. It is possible to feel a relatively different atmosphere from the one expected (and this maybe has become a *cliché*). But what is significant is that this supposed atmosphere of the place is recognized by us in any case, whether in order to adapt to it or to escape it. The atmosphere can, so to speak, be read physiognomically in the external situation to such an extent that we can define it and describe it to others even without being truly touched by it. One might also distinguish between a) the mere conventional knowledge of external atmospheres,[19] b) the felt-bodily and empathic understanding of the atmosphere felt by others, and finally c) the actual and felt-bodily sharing of the present atmosphere. Perhaps this explains why "one can plan to assign atmospheres to an architectural space, but not moods" (Hasse, 2005, p. 370). In general, this may be why we can distinguish – no matter what words we use – between something more objective (the tuned space, the atmospheric), that is to say the objective "roots" of an atmosphere, and something more subjective and ephemeral (the atmosphere in the strict sense of the word, cf. Jäkel, 2013, p. 169 and Böhme, 2001, passim). The resistance to an atmosphere that is in any case perceived may even be conditioned by the inability to let oneself go or by a pathological protest of our state of mind, and could in turn create a qualitatively different and even antithetical atmosphere.[20] This happens when an architectural atmosphere of joyful order is so dull and conventional that, rather than making us rejoice, it ends up making us sad. So, there are certainly atmospheres that arise from the dissonance and discrepancy between what we feel and what we would expect to feel, and even from indifference. But this does not make it legitimate to reduce atmospheres to a mere product of a subject that forgets their own unconscious projection of an expectation over them.[21]

It is important not to overestimate the receptive (cultural, historical and individual) variability. Culture is conceived here not as a collection of mental representations residing in the brain but as something that grows in the felt body, giving rise to a particular responsiveness also to certain atmospheres in the environment. Well, these atmospheres seem to be relatively invariant compared to the different reactions that they can lead to. Let's take just one example that I have already used many times: the impressive entrance hall of a major banking institution will express an aggressive atmosphere of power for those who venture there in search of a loan (whence perhaps the impulse to leave the centre of the room to take refuge in protective nooks and crannies), while subtly expressing a quiet atmosphere of proud belonging for an employee who has developed a strong *esprit de corps* (whence their proud strut in the middle of the hall). And yet what generates both atmospheric moods (a conscious aversion and overwhelming awe or an unnoticed sense of wellbeing and pride) is precisely the "same" spatial-emotional atmospheric quality of solemn and intimidating vastness suggested by whole architectural impression. For obvious reasons, though, the former person fears while the latter loves this vastness and solemnity, which thus also acts as a kind of context or guideline for any subsequent and even contrasting perception. In short, "all people feel these position-qualities overall, even though one feels good and another does not" (Dürckheim, 2005, p. 79).

My thesis is that we experience an atmosphere as if it came from the object only because it actually comes from the object. Like any affordance, the atmospheric one does not change completely

> as the need of the observer changes. The observer may or may not perceive or attend to the affordance, according to his needs, but the affordance, being invariant, is always there to be perceived. An affordance is not bestowed upon an object by a need of an observer and his act of perceiving it. The object offers what it does because it is what it is. [Atmospheres] seem to be perceived directly because they are perceived directly.
>
> (Gibson, 1986, pp. 138–139, cf. also Withagen et al., 2012)

In other words, they never depend entirely on the state of the observer. And the intensity of the protest of our mood in the case of an antagonistic encounter is indeed the best proof of the objective effectiveness of the atmosphere we react to, because it sometimes accentuates our previous and contrary mood, such as when our confusion is exacerbated by the atmosphere of architectural rational orderliness we come across. As mentioned, the prototypical atmosphere is mainly experienced as a first impression. So, the atmosphere of a building or of a city, of their "skin" (Hasse, 2000, p. 168), is not first of all generated by a cumulative cognitive and experiential process but, from the very beginning, counts as a quasi-objective biotopic and psychotopic landmark (no matter how virtual).[22] However, the second (and third, fourth…) impression could give rise to a less intense atmosphere. As a dynamic and chronologically structured process, atmospheric perception could very well change sign and colour for various reasons, especially if we assume that there is an entangled unity between our feeling and sensing, between perception and motor action.[23]

Let's consider some of these reasons. An atmospheric perception may change, for example, because of an even slight change in the perceptual space, but also due to an additional cognition and a deeper appraisal.[24] The (perceived) lightness, refinement and elegance of the London Eye are spoiled, for example, both by the memory of the heavy bulk of concrete that had to be poured into the river to support the structure, and by the thought that a structure born to be transient was made permanent (cf. Rykwert, 2000, p. 265). The discovery of the fictional or even manipulative character of the proposed (architectural) atmosphere reflectively cancels its effect (at least in part), as in Kant's famous example of the "artificial flowers (which can be made so as to look just like natural ones)": as soon as the perceiver finds out "how he had been deceived, the immediate interest which these things previously had for him would at once vanish – though, perhaps, a different interest might intervene in its stead, that, namely, of vanity in decorating his room with them for the eyes of others" (Kant, 2007, pp. 128–129). Similarly to what happens with natural beauty,[25] a wild atmosphere ceases to be such when, sensorially perceived components being equal, its origin is ascertained to be artificial. Sometimes it is instead a new salience that re-tonalizes the whole situation, which was previously anonymous. Or it could be sufficient to experience a variation in the perceptive distance, as when the atmosphere of magnificence of a building resolves, at a close distance, into an atmosphere of decay. Or, even more simply, a change could be caused by a shift in the physiological conditions of the perceiver, or it may also depend on the season, time of day, light conditions, the speed with which one approaches the place, and so on.

Atmospheric perception is therefore at least partly cognitively penetrable and not totally deterministic; it is relatively (perceptually) amendable, even if only on the level of common sense (i.e. of ordinary perception), for example due changes in the perceiver's ability to "note" new aspects. But this always happens within the "same" atmospheric framework of perception, meaning that an atmosphere is filtered through the perceiver's ideas and evaluations and is indeed an invitation that can also be declined. This is a variability that common sense advises us to admit, without thereby embracing projectivistic relativism, by which the same affective space or building could afford completely different behaviours and feelings for different people, and even for the same person at different moments. If by observing an atmosphere, walking in a city or in a building, we *ipso facto* altered it, no first atmosphere could ever overwhelm us and affect us. What happens is, rather, that our first atmospheric perception can change in time on the basis of changed psychogeographical and psychoarchitectural conditions. An initially indeterminate atmosphere can specify itself, centring on a specific and proper anchor point or a different and improper sphere of condensation. The atmosphere changes even more, however, when it is the provisional and fragile outcome of non-homogeneous sub-atmospheres, and perhaps in conflict with the whole (the interiors, for example, could be in conflict with the whole building and some buildings with the whole city). Or when the beholder makes scale-changes in their focus and, for example, what was the figure (apartments, the layout of the rooms or even a stairway) becomes only a background for other entities (buildings, residential areas, cities, etc. cf. Reichhardt, 2009, pp. 52, 56, 77) and so on. This is not too surprising, considering that, just like every other experience in act, atmospheric perception as a multifactorial process imposes its own data (hence the passivity of the synthesis) but still in a context of open possibilities (walking through a building or seeing it from the outside, changing its use, knowing its history or not, etc.).

Summing up, an architectural atmosphere can overwhelm us, it can find us in tune with it, it can be recognised without being really felt, it can elicit a resistance that pushes us to change it (only relatively, given the solidity, in every sense of the term, of a building or a square), it may (for various reasons, also absolutely idiosyncratic) be perceived differently in the course of time, and it may be so dependent on the perceptual (subjective) form or mood as to concretize itself even in materials that normally express completely different moods. At times, the subject-dependence can impede even the mere sensorial-affective observation of the atmosphere one is in, thus causing an embarrassing atmospheric inadequacy for oneself and for others, like when one is moving in a disorderly manner in church or in a too conventional way during a party. In any case, even if atmospheres in the proper sense are prototypical, the atmospheres (or spatial moods, cf. Heß, 2013, p. 58) that we experience most often in everyday life are an "in between", made possible also in architectural space by the (felt-bodily but also social and symbolic) co-presence of subject and object. They are something that exists "between" the object, or rather the environmental *qualia* – including the empty spaces created by the architecture itself – and the subject, or rather their felt body. Nevertheless, this assumption of an inter-corporeal communication (the "in between") must not be reified to the point of counting as "a third element which, as a membrane, is interposed between the two margins" (Schmitz, 2002, p. 71), because this would lead us from the frying pan into the fire, namely to strengthening the dualism that I am seeking to avoid. In a holistic architectural relaxed atmosphere, the relax is indeed not so much *my* relax, but rather and first of all a relaxed situation, in which subject and object are not (yet) independent and isolable parts.[26]

Also, an trivially, an atmospheric situation can actually attract or repel certain feelings and behaviours in an agent if and only if they perceive this atmosphere, and for this reason motivation and intention appear to be needed to explain why they "utilize" certain atmospheric affordances and not others (or filter them in a certain way) at a certain moment in time. Much also depends on the perceiver's ability to detect the atmospheres they encounter, which – I repeat – determine the way we feel in their presence far more than the atmospheres can act as an expressive pretext for the subjective feelings that we may want to project on them. Atmosphericness, as a primary example of expressiveness, is located in a certain pre-dimensional portion of space. This is shown by the fact, very banal but often shamefully underrated, that with the disappearance of a given localized space or building the perceived atmosphere inevitably disappears as well. In fact, I believe it is delusional and subjectivistic to embrace the situationist attempt to redefine architecture as pure atmospherics, to search for passionate trajectories or *ambiances* by means of psycho-geographical *detours* and thus to "redesign architecture itself by exploiting the radical potential of atmospheres" (Wigley, 1998, p. 24).[27] If it's true that I dwell in the space and the space dwells in me, i.e. that the lived space outside and my felt-body supplement and define each other through their resonance (cf. Pallasmaa, 2005, p. 40), it's equally true, whatever the Situationists claim with their creative drifts, that we cannot project any subjective meaning on external forms and gestures.

Finally, it is clear that dealing with the link between architecture and atmospheres means investigating the source of all (atmospheric) problems, that is, the generability (or not) of atmospheric feelings. When presenting their projects, architects often speak of atmospheres. But do they actually feel this "obscure object of desire" (Kretzer, 2013, p. 117) or do they just imagine it? And can an atmospheric *Erlebnis*, with its vagueness, openness and random pattern, be really planned and brought into being through technical constructs, especially today by means of a tool as distancing as the laptop? Shouldn't we think instead that an atmospheric effect is something unavailable (cf. Pinzer, 2012, pp. 107–108)? This would be even more evident if we admitted that atmospheres never exist, if not in a very inappropriate way, as purely potential (virtual, designed) states. There are no recipes, of course, in planning atmospheres. Architectural drawings can do nothing but concentrate "on a specific site [and] try to plumb its depths, its form, its history, and its sensuous qualities", and try thus "to express as accurately as possible the aura of the building in its intended place" (Zumthor, 1998, pp. 36, 13). The architect must then simulate, thanks to a prognostic body-schema and role-play competence, the future body-feeling of the beholder/user, identifying with their situation and therefore perhaps with the desired future atmosphere. Last but not least, architects explicitly rely on a "pact of generosity" (Sartre), on some form of complicity, also as a tacit bodily knowledge, on part of the beholder/user.

Design is certainly always the exploration of a foreign domain, but the claim that atmospheres always "escape analysis" seems greatly exaggerated to me. The same goes for the statement that "any specific proposal for constructing an atmosphere [...] is no longer atmospheric", because "an atmosphere may be the core of architecture but it is a core that cannot simply be addressed or controlled" (Wigley, 1998, p. 27). When it is only planned, an atmosphere of course does not yet exist in the real sense of the word, as it is not yet an event of the real world, but this does not mean that architecture works hard to create nothing but "the illusion that an atmosphere can be controlled" (ibid.). The design of an atmosphere often works perfectly, and if there is an illusion at work in planning, it is the heterogenesis of ends that affects every human activity and that can never be completely avoided. Many urban atmospheres are usually the unconscious and unintentional result of planning, such as in the case of Manhattan, which offered its inhabitants "the spectacle, inscribed in stone, concrete and steel, of a way of life obeying a very different program, one answering a question quite different from that of 'housing'" (Damisch, 2001, p. 110), thus leading to results that were far from expected.

In short: it would be better not to exaggerate the strict actuality of atmospheres. Firstly, they undoubtedly also depend on the co-perception of past and expected atmospheres that are not in act, such as when, for example, the atmosphere of a hospital is tense precisely because we anticipate the situation to follow (the visit, the diagnosis, etc.) and we remember earlier ones (further waits, etc.). Secondly, the atmosphere in act may certainly be also the successful outcome of a design about which we can counterfactually think (what atmosphere would the patients and their family feel if there was a door there, a window there, a view of a garden from the window, etc?). We can indeed imagine, for example, the architectural conditions under which the atmosphere of that hospital could be less tense. An atmosphere thus sometimes exists, obviously in an incomplete and only hypothetical way, also as a potential to invite a certain mood and activity. I do not think that architects do not properly produce atmospheres but only suggest and evoke them in the beholder, and that for this reason they should simply design more neutral places that stimulate the hermeneutic and emotional creativity of the user (cf. Hahn, 2012a, p. 28). In designing buildings architects often do something more precise: they should indeed contrive places that invite certain behaviours, and therefore they should be aware of how to create affordances and how they are perceived (cf. Warren jr., 1995).

It's true that, from another point of view, like moods or *Stimmungen*, atmospheres are sometimes more of a transcendental unconscious or a background perceptive condition than a thetic object of transitive perception (cf. Thibaud, 2003, p. 293). They are sometimes the generic condition of specific experience – what simply allows "emotions to emerge, to be" (Zumthor, 1998, p. 27) – rather than a specific and focused experience. But this is not always the case, because at other times the atmospheric encounter can to some extent be perfectly planned. Far from finding in people and things only generic and vague occasions to manifest themselves, atmospheres are sometimes precisely the expressive way in which things (in our case architectural places and "things") call for us or even look at us,[28] being much more than a simple gaze-of-response. Even a building can look at me and atmospherically make me, as Sartre would put it, "no longer master of the situation"! I repeat: the perception of what a situative atmosphere affords, also (if not emblematically) in the case of architecture, despite being only an opportunity and not a cause, should never be confused with the colouring of experience by subjective needs and motives.

When atmospheres solicit a certain feeling and activity, the agent/perceiver feels immediately drawn to feel and even to act in a certain way. Many other issues, of course, deserve to be further discussed, starting from whether architecture today, in a culture of cynical "homelessness" (nomadism etc.) encouraging a sort of deleterious atmosphere-zapping, should generate atmospheres only in order to achieve some other purpose or fully autotelic atmospheres. However, for now suffice it to say that we should steer away from neo-Romantic regressive temptations, which are never totally absent in the atmospherological discourse. There is indeed an architectural atmosphere not only when we recognise an organic situational reunification, but also when we experience an irritating lack of specific affective-value cores in certain places: perhaps, and unfortunately, even the inhospitality of the contemporary city is an atmosphere, under the guise of a "petrified nightmare" (Mitscherlich, 1965, p. 22). However, we must also avoid the opposite theory, according to which we can consider aesthetic only the exposure to foreign and provocative spaces, which for this reason should be more subject to creative interpretation (cf. Haapala, 2004, pp. 49, 51).

In short, we must avoid both these extreme approaches and thus neither identify an atmosphere only in an organic and identitarian form[29] nor see it as the result of subversive designs or practices. In other words, in line with my purpose of de-axiologising the notion of atmosphere, one has to reaffirm that even the negative emotional and felt-bodily involvement that we feel in the most chaotic and anonymous city is something atmospheric. And what we feel is atmospheric also when we are in cold and abstract buildings, even in those anonymous non-places that we could, by analogy, define "non-buildings" and that are totally devoid of the "gentleness" that one might rightly expect from good architecture.[30] If almost everything we know or sense is conditioned by atmospheric affordances, even the very aim of the "rejection of atmospheres constructs a particular atmosphere", such as when planning "is meant to be subordinate to reason, controlled by the line" (Wigley, 1998, pp. 26–27).

If it's true that architecture is its own time apprehended in buildings (cf. Böhme, 2006, p. 174), its atmospheres will be indeed as authentic as the general needs and moods of the time. This could easily bring us from aesthetics to politics. But this is not really the place to determine whether a good political solution, as I argue, is to design an architecture that allows for a quick alternation between an uncritical-pathic immersion and a more critical-rational emersion.[31] Rather, the time has come to move from the atmospheric nature of architecture in general to the very special atmosphericness of dwelling.

Notes

1 A tradition that spans from the 19th-century "Carpenter law" to contemporary mirror neurons.
2 It's much more difficult to accept the idea (Schmitz) that the atmosphere is also fully independent of the senses.
3 "The meaning or the value of a thing seems to be perceived just as immediately as its colour" (Gibson, 1986, p. 138).
4 That means, according to Straus, that we feel ourselves and the world, ourselves in the world and ourselves with the world.
5 The way I make something and how I feel about it have an influence on how a situation appears to me, making this also an aesthetic question (see Kamleithner, 2009, p. 1).
6 Norman (2005, p. 10, p. 13), reminds us that "pleasing objects enable you to work better […] When you wash and polish your car, doesn't it seem to drive better? […] When you bathe and dress up in clean, fancy clothes, don't you feel better? Emotions are judgmental and prepare the body accordingly […] Cognition interprets and understands the world around you, while emotions allow you to make quick decisions about it. Usually, you react emotionally to a situation before you assess it cognitively, since survival is more important than understanding".
7 For recent analyses of the relationship between architecture and atmospheres see Tidwell (ed., 2014) and Pérez-Gómez (2016).
8 He also means the "sound of silence", because architecture (as petrified silence) must maintain and spread the silence against the abuses of communication.
9 Even through their "physical and in all likelihood psychic" temperature (Zumthor, 2006, p. 35).
10 Dwelling is the culture of feelings in a closed space. "The spatiality of feelings becomes here so efficacious that human beings always need these places of dwelling – instead of a simple sheltered accommodations – in order to capture feelings, in a way, administer them and model them on these places". In a nutshell, "dwelling is managing the atmospheric" (Schmitz, 1990, pp. 318, 320)
11 Of the felt body itself and not only of a body's idea (cf. Zumthor, 2006, p. 23). "Buildings must be designed starting from the lived body" (cf. Waldenfels, 1996).
12 A sort of sculpture made dynamic (cf. Jaschko, 2011, p. 113).
13 According to Jäkel (2013, pp. 186–188), they are: space creation, space articulation, quality linking, expressing moods, dynamizing, motivating expression, inseparability of aesthetic and functional execution.
14 Especially today, when numerous architects are attempting "to re-sensualise architecture through a strengthened sense of materiality and hapticity, texture and weight, density of space and materialised light" (Pallasmaa, 2005, p. 37).
15 For an analysis of Zumthor's work in these terms cf. Heß (2013).
16 A description that Jäkel (2013, pp. 178–180) summed up in five stages: 1) occupancy of a space, 2) forms, 3) movement and action pattern, 4) physiognomic dispositions, 5) gesture exchange. She also applies these criteria (always a bit too vaguely, I believe) to three houses designed by Lois Welzenbacher (ibid., 193 ff.).
17 I am obviously referring mainly to the vector atmospheres, namely the atmospheres that correspond to the directional space (cf. Griffero, 2014b, p. 46).
18 "Architecture is to be found in the relationship between atmospheres, the play between microclimates. The encounter of these seemingly ephemeral atmospheres can be as solid as any building" (Wigley, 1998, p. 24).
19 But we can only seldom legitimately refer the objectual immanence of the expressive atmospheric characters to their "huge social invariance" and the contrast between personal feeling and perceived atmosphere to that between "different

levels of subjectivity", particularly between how I should feel something and how I really feel it (Hauskeller, 1995, pp. 29–30, pp. 45–46).

20 For Hellpach (1939, p. 120, fn. 48) "we all know from experience that our melancholy can be dispersed by the serenity of the environment (by irradiation), but it can also be re-sharpened by contrast".

21 As argued by Hauskeller (1995, p. 46): "It is therefore not an objective expression of the landscape that is opposed to subjective feeling, but it is the way I should perceive the landscape (i.e., as serene, having previously perceived it this way) that is set against the way I actually perceive it (i.e. as sad)"

22 See Griffero (2013). There are city views known only through cinema, such as the Manhattan skyline seen from normally impossible frontal or aerial perspectives, or through a kind of iconic synthesis (Moscow=austerity and hesitation, Las Vegas=hallucinated spectacle, Los Angeles=web of infinite flows): see Thibaud and Thomas (2004).

23 It is the "ring-shape"model (Viktor von Weizsäcker's *Gestaltkreis*), according to which the action exerted by the things and by the surrounding space change the attitude of the beholder and this in turn modifies the orientation of their perception.

24 This is a more reflective and indirect stage that, accordingly, should not be considered more aesthetic than atmospheric (as Reichhardt, 2009, passim, wrongly does): precisely because it's a largely non-cognitive state of mind but a peculiar "reasoning feeling", the first impression is already an aesthetic situation.

25 "The beauty should be that of nature, and it vanishes completely as soon as we are conscious of having been deceived, and that it is only the work of art" (Kant, 2007, p. 131).

26 As Bollnow (2011, pp. 230–231), says in regard to landscape, "there is a mood in both the human soul and a landscape or an enclosed internal space, and the two things are in principle only two aspects of one unitary being-tuned […] The mood itself is not something subjective 'in' the person nor something objective – as something that could be found 'outside' in the environment – but it regards the subject in their yet undivided unity with their *Umwelt*".

27 I am not convinced of the total metamorphic power that the situationist art of the *dérive* has over the urban ambiance (cf. instead Vazquez, 2010).

28 This is true not only, as Sartre claims, with regard to the atmospheric irradiation of the gaze of others, but also for the appearance of a thing (cf. Griffero, 2017a, pp. 93–101 and Jäkel, 2013, p. 94 ff.).

29 Architecture must obviously generate some habitability, i.e. it must represent an appropriate framework for our daily lives, even if only in the sense of resilience as the inner individual's ability to properly adapt to stress and adversity (Graf, 2006, p. 1).

30 This does not mean (as Pallasmaa argues, 2005, p. 19) that one could not fall in love there!

31 See Bieger (2011, p. 93) and Griffero (2014c). Only an adequate atmospheric competence (both productive and receptive) could immunize us against the media-emotional manipulation produced by the aestheticization of politics and social life in the late-capitalistic "scenic" economy.

10 A house is not a home

Dwelling means cultivating atmospheres

A pathic and atmospherological aesthetics is critical of nostalgic deferrals to the past, utopian deferrals to the future and ontological deferrals to the non-simply-present. In it, as we have already seen (Chapter 8), presence returns to forcefully assert its rights – not least also as a dwelling presence. One moves around and works in the city, one lives in it and feels, as we have seen, its atmospheric "skin". However, it is especially at home that one really dwells (in a felt-bodily way), as prototypically shown by the overcoming of nomadism already in the figure of Ulysses (before Dante's representation of him). Unfortunately, dwelling is often mistakenly understood not as a fundamental pathic and cyclical condition[1] but as a merely pragmatic (social, financial, etc.) process (cf. Hasse, 2009, p. 221), or even, because of a presumptuous allergy to every "border", as a simple attachment of the virtual (cf. Zaborowski, 2008, p. 195). Instead, dwelling demands once more philosophical attention. Once again we must learn how to dwell, as famously suggested half a century ago by Heidegger, who saw building and dwelling (somewhat vaguely, I think) as taking care of the Fourfold (earth, sky, gods, mortals). One must ask, however, not so much whether the human being can reach their true essence only in dwelling,[2] nor – with a metaphor that correctly reaffirms the pre-intentional ontological intimacy of humans with the world – whether we also inhabit time, the world and things, as posited by the later Merleau-Ponty. Rather, more simply, one should ask what a "true" dwelling is from the atmospherological point of view. And the answer is clear: it is the condition (Bollnow, 2011, pp. 288–289) which, while preserving us from a foreign and hostile space by virtue of the presence of affectively dense things and a non-phobic lived space, also prevents us from ending up rigidly encapsulted in a cement envelope, completely blind to the outside world.

Of course there is much phenomenological evidence of the affective centrality of dwelling. We know that sometimes a room is enough to change the way we feel, regardless of whether we perceive a uniquely positive atmosphere (brightness) or a bittersweet one, like when, for example, we anticipate its inevitable future decadence (cf. De Botton, 2006, Chapter I). We also know that a homeless person is not only someone who has no roof over their head or four walls around them, and that only a snail can feel at home everywhere,[3] just as know that it is not good to walk into someone's house, without knocking or calling first, only because "the door was open". It is our lifeworld itself that shows it: dwelling has never ceased to be rooted in a non-Euclidean (anisotropic) and safety space. This sphere coincides with our identity[4] to the point that saying "come over to mine" is equivalent, in fact, to saying "come to my house", which is an invitation to access "an organic unity, whose essence is some definite power" (van der Leeuw et al., 1955, p. 395), valorizing the present and presence. Despite being historically and culturally variable from a proxemic point of view, dwelling always presupposes a "not too large" size which seems to make a certain well-being possible, in other words an atmospherically significant situation that is irreducible to the sum of its parts and which consists in "inscribing the things and places of one's environment, neutral in their meaning, into a profile of personal meaning" (Hasse, 2008, p. 109).

Of course, with the advent of the modern metropolis, this meaning too often becomes only the mnestic-nostalgic trace of the (presumed) idyllic condition of small rural centers, that is, a lived space – hence a place in an anthropologically identitarian sense – which would be as closely intertwined with the person as the animal with its territory. Indeed, there is always some *laudator temporis acti*, scared by the fact that "at once the ceiling is full of leaks, the walls become transparent, the cement uniting the members of the family crumbles away, the shared privacy disintegrates" (Anders, 1956, p. 17) because of the extreme media porosity of our homes. Evidently the nostalgic, not unlike the archaic caveman, person is convinced that only in the sort of "third skin" of dwelling – an authentic physical and psychological refuge[5] from external disorder and opacity – can we be truly ourselves, that is, "tell" our authentic personality, thus staging (even for ourselves) a ritual and felt-bodily process that becomes a real "psychic landscape".[6] Nostalgic from head to toe, they may even claim, suggestively, that "in Paris there are no houses, and the inhabitants of the big city live in superimposed boxes" (Bachelard, 1994, p. 26), in that, where there are no cellars (and therefore roots), no heroic verticality (inhibited by the elevator) and no cosmicity (due to detachment from nature), an anti-dreamlike horizontality allegedly prevails. The nostalgic would also add that it is not possible "to live" in a hotel or in a nomadic way, or that it takes two to truly dwell, and that one needs a female presence in the house (Minkowski, 1954, p. 180 ff.; Bollnow, 2011, p. 146), so much so that a simply furnished flat or the apartment of a bachelor or widower can hardly become an intimate interior. Or again, the nostalgic might say that one really dwells only when one can count on that fraternal proximity that atmospherically pervades things and places.

Well, this not-so-hypothetical nostalgic is perhaps overestimating and eternizing the family, concealing the undeniable social changes that have taken place and forgetting that the transition from the public theatricality of pre-modern living, both aristocratic and plebeian, to the intimacy of living privileged by bourgeois society is actually quite recent. Perhaps they are losing sight of today's irreversible artificialization of living: rather than guaranteeing a defense from disorientation, in the name of an obsession that is no longer moral but functional (Baudrillard, 2002, p. 29), it imposes the ergonomic conformation to the imperative of efficient housekeeping, and in any case it relies on objects of lower transitional value due to their inexhaustible availability and replaceability. In short, because of the loss of the experience of places – perhaps due to the leveling of the (even non-political) forms of life – the intimacy of living has perhaps really become a pale larva, reduced to the container of an external (and past) world that is excessively close (or far) and in any case has been derealized, with respect to which the family would have turned into a sort of miniature public. Or perhaps dwelling can still be a stage that, precisely in its decorative values (in the positive sense of the term), acts as a psychological background for the expressive personality, more individual than social, of the resident-character (Sparke, 2008, p. 92). In a word: is dwelling still a biotope and an indispensable psychotope? Or has it long ago turned into a mere surrogate identity, in a jargon that claims to be an original (habitative) idiolect? Besides, the latter is largely affected both by a professionalising minimalism, which relies on a few efficient steps according to the principles of modernist ergonomic rationalism, and by the frequent inclination of progressive architecture to "erode the boundaries between the inside and outside and to enhance the porosity of their architectural structures" (ibid., p. 167), thus reaching the – undesirable, I think – interior transparency, understood as an object of art to be contemplated rather than to be lived.

The ontological question, then, turns into a historical-cultural question. When is it that the pleasure of relaxing between four walls has ceased to appear essential and natural? When did it become pseudo-homeliness and assumed the traits of the "horror of comfort" and of the "secretly vexatious" (Mitscherlich, 1965, p. 11, p. 125)? A time came when we sought to compensate for the ugliness of the surrounding landscape, through a form of resistance to the new that was able to receive the whole world at home (at least in effigy), up to making us "boon companions of everything on our planet and in the universe" (Anders, 1956, p. 23). Perhaps only then did dwelling turn from an existential principle into a culturally conservative and psychically regressive alibi. Or perhaps this process has been going on for much longer, at least since the 19th-century culture, by enveloping the Victorian individual into a "compass case", turned the living room into a "box in the theatre of the world" (Benjamin, 2002, I, p. 220, p. 19) and, at the same time, into a shelter full of artifacts, specifically female due to the increasing externalization of male labour. In short: is it really only with the advent of radio, TV and the internet, that we started to talk and look at each other, even at home, only through a third party?

To give an answer to these questions is anything but easy. If only because the metaphysical extraterritoriality, already implicit in the super-terrestrial and ubiquitous orientation of the major monotheisms,[7] and then secularized in the Enlightenment as well as in its technoscientific consequences (if you stop you lose!), seems to be decidedly in crisis today. The suggestion of nomadism, both axiological and digital, seems increasingly weakened by anxieties of identity and locality that were previously dormant, and which it would be absurd to disregard (if only to adequately deal with them). On the one hand, the insularism of Western dwelling (maximally symbolized by the height of the skyscraper), resulting from the privatization and subjectivation of an existence that was profoundly "incarnated" in one's home,[8] is so powerful as to demand a (far from obvious) architectural design starting from the interior. On the other hand, it appears to be radically questioned by the *flâneur*: not so much the metropolitan one, for whom the streets were ultimately equivalent to the four walls of a house – a model that basically disappeared at the end of the *belle époque* (Schmitz, 2008a, p. 33) – but the digital one.

Contextually, this insularism is questioned by a sort of externalization of interiors, "transformed into an object of mass consumption" (Sparke, 2008, p. 55), able to guarantee – sometimes even in public places – an immersiveness similar to that hitherto only provided by private homes. The Western necessity of a private and reflective individual sphere, functional to the construction of both one's self and one's social status, justified two things: in terms of design, the birth of domestic interiors strictly separated from the interior dedicated to public activities; in terms of geographic-architectural theory, the obvious and therefore generically reiterated value opposition between interiority and exteriority (Relph, 1976, pp. 49–55). Well, this necessity was soon eroded by the so-called "homes from home" (luxury shops and lobbies of big hotels, first-class carriages and clubs, today pedestrianized areas and shopping malls, etc.), which provided an idealized replica, though much less private and reflective, of domestic spaces. Yet even in modernism, at least in the "humanized" one, efforts have been made to recover the previously stigmatized need for intimacy, that is, a clear division between the private-reflective sphere of dwelling and the anonymous-guarded one typical of public spaces, so that dwelling could be (once again) a real "material score for sentimental events" (Meisenheimer, 2008, p. 43),[9] to such an extent that memories seem to belong not to me but to the places themselves and are destined to survive when I am gone (cf. Minkowski, 1954).

And so, relatively freed from the sense of guilt coming from nostalgia, let's consider once again personal dwelling. Far from being an only representative staging, it is never given by "a constellation of things and of relational arrangements" but by an "overall context of meaning" (Hasse, 2009, p. 180) able to release specific socio- and psycho-environmental *Stimmungen*.[10] What I want to talk about here, though, is not so much the specific articulation of this domosphere, in the continuous and productive tension between the abstract choices of architectural planning and the affective ones, always *in fieri*, of the inhabitant – first simply functional equipment, then representative furnishing, and finally personal decoration. Rather, I wish to address the aesthesiological-atmospheric *Uphänomen* of dwelling as a symbolic link between space-time and the affective sphere. According to Hermann

Schmitz – a true *spiritus rector* (though not from the strictly aesthesiological standpoint) of my pathic aesthetics – dwelling, as already mentioned (Chapter 1), is a form of culture-cultivation of feelings within a defined space[11] which is sacred (broadly speaking), be it a home, a church, a garden,[12] or a Japanese tea house.[13] This practice "produces a relationship with and between places and thus constitutes spaces of feeling". It is an "imaginary cartography of affective vital significations" (Hasse, 2008, pp. 109–110), but only insofar as it generates a temporary balance with atmospheric feelings that are otherwise abysmal and therefore potentially lethal. Sometimes dwelling is able to displace these feelings into interstitial spaces that are not strictly functional, as Tadao Ando claims, or to condense them in the most symbolic places of the house: in any case, expressing the primary need to domesticate feelings, dwelling is certainly not an invention of modern bourgeois intimism, fossilized on the (dubious) virtues of the small things of everyday life, but the indispensable need for fine sensoriality and for a controlled atmospheric pathicity.

It is not enough – I repeat – to have a "roof over our heads" to say that we are really dwelling somewhere. It is necessary to have a place and a situation that, by shielding a reality which, however, we obviously do not want to completely ignore, allow us to capture, cultivate and administer external atmospheres otherwise out of control (in this sense, as mentioned, dwelling is like art and law). Managing and filtering the pathic in a protected space, dwelling gives life, therefore, to an intense and nuanced affective climate able to mitigate the external uncanny, which nowadays has certainly been delocalized following modern mobility, and has maybe moved from the forest to the urban "asphalt jungle" (Schmitz, 1977, p. 245; Soentgen, 1998, p. 83), or even to cosmic space, but which anyway is still very much present. Think, for example, of the lively creaking of wood in the fireplace in a country house, or the maternal noise of dishes being handled for a child struggling to fall asleep, or of the tactile sensation of the softness of a sofa, or of the penumbra generated by the interaction between light-points and furniture drop shadows: the atmosphere cultivated by dwelling seems to be, in these and other (perhaps even less obvious) cases, the relaxing familiarity that comes with finding all things in their right place. That is to say, in "surroundings" that, according to the externalizing and anti-psychological orientation of atmospherology, express these (usually friendly)[14] atmospheres autonomously, instead of being their projection plane (cf. also Guzzoni, 1999, p. 33). After all, we can easily see this for ourselves: when moving to a new house, we typically "feel" for the first time that a piece of furniture to which we were previously indifferent is actually indispensable.

Let's now try to apply my distinction between prototypical, derivative and spurious atmospheres to dwelling. a) For example, consider structural changes capable of generating a different objective atmosphere, as in the case of different criteria to arrange rooms: rooms used to be linked to each other (a "Grand Central Station atmosphere", Hall, 1966, p. 104), then were placed around a corridor or atrium (whence a more stable domestic atmosphere), and finally today they are dissolved into large open spaces, with their (illusory) atmosphere of vastness and democratic transparency. b) Consider instead less impactful aesthetic choices, which can generate residential satisfaction but also undesirable atmospheres, such as when excessively elegant furniture ends up inhibiting conversation, domestic hygiene turns into authentic fetishism, or the privilege of the hall as a disproportionate representative space is equivalent to exhibiting a "stately room with no castle behind it" (Mitscherlich, 1965, p. 138). c) Finally, consider the atmosphere of narrowness that is idiosyncratically privileged by the child for its enveloping and protective character, while being felt as suffocating by all adults, albeit because of the same felt-bodily resonance.

For me, dwelling is first and foremost the search for the (aesthesiologically and pathically) right atmosphere in an interior: an atmosphere is the right one because, by virtue of its synaesthetic characters (although relatively variable),[15] it protects a home's privacy, comfort and warmth, but can also satisfy our socio-expressive needs, possibly also addressed to the outside world. But a dwelling atmosphere is right, that is, comfortable and intimate, not so much and not only when, as in Nordic culture, it aims to replicate on the residential level the aesthetic utopia of the *Gesamtkunstwerk*, but, more simply, when it becomes for a felt body an indispensable and exclusive place of self-representation: "to curl up belongs to the phenomenology of the verb to inhabit, and only those who have learned to do so can inhabit with intensity" (Bachelard, 1994, p. xxxviii). It is easy for traditionalists like Bachelard, Minkowski and Bollnow to defend the (for them) necessary transformation of dwelling in a (positive) solid psychic basis: acting as nest, shell, cottage, crypt, organic habit, centre of reveries, oneiric verticality (psychic and therefore also cosmic), sophisticated miniature of the cosmos, and mnestic-oneiric image, in fact, "a nest-house is never young" (ibid., p. 99). Rejecting the obsession of existentialist gnosticism with the fear of primal disorientation, it seems to effectively testify to the opposite idea, namely that "all of life is well-being. Being starts with well-being" (ibid., p. 104). Indeed, with dwelling and its indispensable atmospheres.

At this point, I must stress the already mentioned circularity of dwelling, i. e. the need to have an interior in whose safety and transitional objects we can cyclically withdraw, thus abandoning the (more) anxiogenic and exposed external world, perhaps even – yet without Descartes' famous (or infamous) masonry stove – disembodying ourselves from the world in view of a reflection pushed up to the pure *cogito*. Of course, once again, one shouldn't disregard cultural differences. The traditional Japanese house, for example, shows a smoother relationship between inside and outside than the Western one, and there are also cultures in which excessively segregated houses are even offensive to the community. Nor should one disregard gender differences: it is difficult to deny the predominantly feminine value of interior living (albeit politically incorrect, the topos of the "angel in the house" must have some historical-cultural foundation!), diametrically opposed to the exteriority of male dwelling, to the point that a man can sometimes be a "guest" even in his own home. And of course there are also historical differences. Think, for example, of the metamorphosis – plastically exhibited countless times in the artistic field – undergone by typically bourgeois dwelling intimacy, which was transformed into a more uncanny atmosphere than the external one. In fact, in the 19th century, chaotic and threatening external daemons – whose favour the Romans tried to obtain by bringing them into their homes – were replaced by haunted houses, i.e. by apparently welcoming interiors that, instead, allowing for the re-emergence of what should remain hidden, appeared to be heavily infested by evil forces (cf. Vidler, 1992, p. 36).

Yet, despite all these caveats and the thousand possible exceptions, dwelling, as a lived centre of the world, and therefore as *imago mundi* opposed to the growing secularization and (scientific) peripheralization of the terrestrial space (Bollnow, 2011, pp. 119, 141), continues to guarantee the atmosphere of familiarity that is necessary for the human being to escape the feeling of uprooting, yet without being content with anonymous concrete-reinforced ant-hills. And dwelling guarantees this familiar atmosphere, without turning into a form of desperate segregation, also thanks to the already mentioned vital cyclicality, in the light of which, usually, every day one leaves home, putting on the social mask, but then one comes back to it, stripping away both social tinsels and public clothes, especially in areas dedicated to hygiene and rest (cf. Vitta, 2008, pp. 156–157). Thus, with the noble gesture of setting boundaries – think of how fatal any crossing of the city *pomerium* was for the Romans – in order to cross them, dwelling becomes a fundamental core of human existence, a place so different from others that all the directions of those who reside in it lead there, but which also attracts the happy few for whom the door, as they say, is always open. A salient place from the existential as well as phenomenological standpoint – even for those who, a bit exaggeratingly, feel they are perennially "on the road"[16] – dwelling therefore is the parameter of every closeness and/or distance. In particular, what acts as a parameter is its further interior, that sort of home in the home or "centre of the centre" that acts as an authentic quasi-altar, identified, depending on the ages and cultures, in the fireplace or in the kitchen, in the dining table or in the living room, in the bed[17] – perhaps further segregated by curtains – or in the pudically inaccessible so-called night-zone.

In short, dwelling is the main condition of possibility of our full realization as human beings, the only place where one can truly be oneself. Which easily explains the thrill we feel when we walk into somebody else's house while unexpected, which is as sacrilegious as when we read someone's private correspondence. It is certainly true that one also dwells in a building, in a street, in a city, even in a country, etc., that is, in spaces that are not completely foreign and unsafe just because they are physically outside one's home; and yet it is only when one can rely on a shelter, on protective walls that clearly separate the inner-private space from the external-public space, that one actually dwells in a condition of security (*Geborgenheit*). The latter, indeed, is a phylogenetically and ontogenetically happy condition,[18] because it is nurtured by a "place tonalized by familiarity" (Hasse, 2009, p. 25) – in short, by the harmonious atmosphere of family trust which has been rightly seen as the guarantee of a healthy psychic life.[19] It is precisely this primary, serene and trusting condition that is recalled both by Bachelard, in his poetic-oneiric topophilia, and by Bollnow, more soberly interested in curbing the anguished existentialist vision of a nomadic human condition (cf. e.g. Bollnow, 2011, pp. 126, 132–133). Besides, these authors are both convinced that one can feel at home in ever-new outdoor spaces only if and because one has had and lived a home, with the atmosphericness it implies (or implied).

What is clear, and absolutely decisive in a pathic philosophy, is that the protection offered by dwelling, namely a residential atmosphere[20] different from that of other lived spaces – a university classroom, for example, should obviously encourage concentration instead of relaxation and familiarity – is never reached through rational planning. Rather, on an aesthesiological level, it is reached through a lived and ever-changing integration between the more original, that is, felt-bodily and affective (directional and vast) space, and the relative-local space of surfaces, which does not actually harbour feelings (as posits Schmitz, 1977, p. 237) but can thingly condense their power in various ways. It is not enough to wearily reaffirm the anodyne "real estate" criterion according to which the intimacy of a home depends entirely on its ability to express the personality of those who live there. More concretely, it is necessary to verify phenomenologically whether a certain dwelling generates a pleasant impression of isolation thanks to several aspects: tents, windows that are not excessively large,[21] a non-excessive size (a factor that obviously changes depending on the dweller's lifestyle), furnishing that is neither scarce nor packed, a certain warmth (polymodally understood), the chromatic choices, a middle way between artificial order and a total mess, a "fusion" between person and furnishings, or the "lived" nature of the objects in it that acquire, thanks also to their temporal depth, almost a kind of plant-like growth. Objects and furnishings thus assume a patina that seems absolutely necessary for a home to be comfortable and intimate. But let's move on to an even more concrete example and enter inside the home, in its inclusive space.

Notes

1 A cyclicity (going out and coming back) that is both daily and long term, like when we go back to our hometown for happy or sad occasions.
2 This *leitmotiv* guides Bollnow (2011, p. 126).
3 But for a sociological-phenomenological analysis of marginal homes (prison, convent, ship, retirement home), cf. Hasse (2009).
4 This is even more true given the increase in free time, so much so that one could even speak of a "tyranny of intimacy" (Sennett) or of "immune system" (Sloterdijk).

5 We need a home in the psychological sense as much as we need one in the physical: to compensate for a vulnerability. We need a refuge to shore up our states of mind, because so much of the world is opposed to our allegiances. We need our rooms to align us to desirable versions of ourselves and to keep alive the important, evanescent sides of us. (De Botton, 2006, p. 107)

6 A landscape that is doubly private, as it is specific to the family and also, with the specialization of the living quarters, of each of its components.

7 So far from the idea of a stantiality of the divine (see Chapter 12) that the faithful ends up feeling at home, paradoxically, only when they are elsewhere, for example in pilgrimage.

8 Bollnow (2011, p. 272 ff.) insists on this point.

9 What we do in our homes always implies a reconfiguration of our own self, because "moving, ordering, renovating, and changing furniture are activities that act and feedback intensely on our emotional state" (Funke, 2006, p. 20 f.)

10 Which Baudrillard (2002, p. 30 ff.) investigates in terms of *ambiance*.

11 "The spatiality of feelings becomes so effective here that, again and again, people need these places of dwelling – rather than a simple safe accommodation – in order to capture their feelings, administer them and shape them on these places". In short, "dwelling means having atmospheres at one's disposal" (Schmitz, 1990, pp. 318, 320).

12 Which, incidentally, is the only place (as recalled by Schmitz, 1977, p. 305, pp. 292–299) that is "dwelt" without any other more precise purposes, such as eating and sleeping (home), isolating oneself from noise (tea house), looking after one's body (bathroom) and worshiping God (temple), but simply for the sake of dwelling. This is true for the garden in its original form of *locus amoenus* or *hortus conclusus* but also applies to the apparently borderless version of the so-called garden-landscape. For Schmitz, the latter still implies a "frame", namely the "invisible segregation" provided by a gaze that, ever since the baroque age, has been educated by landscape painting.

13 The ritual takes place in a circumscribed environment, which is the more silent and intimate the more it is the result of an introversion that intentionally contrasts with the (perhaps inclement) weather, but is unintentionally syntonic with other people thanks to an atmosphere of happy mediacy.

14 As Bachelard would say for every piece of furniture (1994, p. 71).

15 For an American and a German, for example, "whether the door is open or shut, it is not going to mean the same thing" (Hall, 1966, p. 137).

16 Bollnow, in re-evaluating dwelling as the fulcrum of living space, explicitly opposes the existentialist tendency to underestimate it as the place of "thrownness", as such in no way able to compensate for the uprooting of life *qua talis*.

17 In which, not coincidentally, our days and even our lives begin and end.

18 Respectively: the archaic and uncrossable magic circle and the small refuge always loved by children.

19 This is notoriously the thesis of Tellenbach's book on the oral-sensory atmosphere (1968).

20 Understood as the (multidimensional) gratification of living in a specific place.

21 By preventing the dwelling from turning into a place of segregation, doors and windows still mark the human ability to both set limits and transcend them (which is Simmel's, not just aesthetological, *leitmotiv*). Allowing one to look outward without being seen, windows, which not coincidentally were originally eye-shaped, represent a device that takes randomness away from what it frames, thereby idealizing or eternalizing a certain portion of the exterior (Bollnow, 2011, pp. 153–155 with reference to Rilke).

11 Inclusiveness and immersion

When we say, far from metaphorically, that a certain room is oppressive or relaxing, or that a very agile movement appears faster than one which actually covers more meters per second (Schmitz, 1990, p. 287), we are obviously referring to phenomena that are spatial but not in the local and physical-geometric sense. These phenomena, indeed, refer to the "absolute" (lived, pre-dimensional) space which the subject corresponds to and participates in thanks to their felt body, by being "pathically" exposed to cross-modal "ecological" suggestions. It is obviously only in this kind of space that it makes sense to talk about inclusiveness. In fact, the allocentric isotropic space formed by only relative localizations cannot be inclusive, nor can it irradiate other affective *qualia*. Only the egocentric or existential space can do so, being anisotropic and made up of absolute localizations, on the basis of which the lived body (*Leib*) perceives and acts as a true zero-point (Husserl). In the light of the increasingly evident neuronal plasticity, even affective neuroscience admits this, speaking of "embodied simulation".[1] Given that the dynamic of consciousness is outside of our mind – since "to have a mind [...] requires more than a brain" (cf. Noe, 2009, p. 10 and Pérez-Gómez, 2016, p. 22) – it is easy to argue that architecture is the artistic field that most relies on the unaware multisensory and felt-bodily contact with the atmospheric effects of our surroundings (Mallgrave, 2013, p. 109).

To speak of inclusiveness means to abandon the physicalistic and functionalist topology and embrace the qualitative one, considering proprioceptive sensations that, while presiding over ordinary spatial experience, are hard to thematize, like "the darkness needed in the theatre to show up the performance" (Merleau-Ponty, 2002, p. 115). In so doing, the qualitative approach, in fact, rejects the physicalistic conception of space as well as the plane geometry that has been its structural presupposition since the age of the Greeks, (re)discovering that "the spectrum of spatiality is much more extensive than what is suggested by the classical conceptualization in the philosophical and scientific field" (Schmitz, 1967, p. xvi). In line with the spatial turn attested in the humanities today, this results in the revival of the phenomenological-psychopathological (Heidegger, Binswanger, Minkowski, Straus and above all von Dürckheim) and anthropological-existential (Merleau-Ponty, Bachelard,

Bollnow) theme of the so-called lived space (Cf. Griffero, 2014a, pp. 36–47; 2014b and Hasse, 2005, pp. 150–198). Only in the context of an atmospheric qualitative approach to the environment can one think, for example, of a "lived distance" that, "besides the physical and geometrical distance which stands between myself and all things [...] binds me to things which count and exist for me, and links them to each other. This distance measures the 'scope' of my life at every moment" (Merleau-Ponty, 2002, p. 333). It is therefore legitimate to say that there is not enough space, that we want our own space, that we need it, that there is too much of it so that we seek shelter in narrowness, that we can or cannot have it, that we can make it (thus creating a void that was not there before), and so forth.

However, if inclusiveness is an atmospheric qualification, it is first of all necessary to clarify, in extreme synthesis, what stratified neo-phenomenological topology underlies the present approach. Local space, in which our entire existence seems to unfold, consists of pairs of points that can be freely constructed, cancelled or modified, always inter-defined by position and distance (therefore always only relative), and marked by an equally relative (and essentially circular) distinction between movement and immobility. Moreover, it springs from the inhibition exercised by some obstacle on the felt-bodily fundamental directions, irreversibly oriented from tightness to wideness, and from the consequent generation of surfaces. Of course, only local space allows for the unfolding of the subjective combinatorial-planning fantasy, turning the felt body into a physical body and directional space into a more abstract-idealized multidimensional space (angles, points, lines, surfaces, bodies). This leads to perhaps even onto- and phylogenetically adaptive advantages such as the optical centring of space. However, local space is always parasitic of directional space, which as such is more original and definitely more salient in our lifeworld.

Inclusiveness plays a small (though not inexistent, as we shall see) role in the external local space made of surfaces. It plays a much greater role in non- or pre-dimensional volumetricity (the space of liquids, sound and silence, of gestural articulation, etc.) whose affective suggestions (anguish, pain, pleasure, freshness, exhaustion, etc.) are perceived not in the physical body organs (*Körper*) but in the various "isles" of the felt body (*Leib*, cf. Griffero, 2017a, pp. 55–77; 2017b; 2017c). In this peculiar voluminousness, based on an involving and quasi-thingly directionality whose anisotropic operator (so to speak) is the "felt-bodily presence", one can indeed feel included or not. On the other hand, in the absence of "a structure of the space of lived bodiliness determined by anguish, vastness and direction" – and, why not, also inclusiveness – "the whole phenomenon of space would dissolve into the intellectualistic haze" (Schmitz, 1967, p. 31). In other words, it would not be the case that "by feeling in a certain way, we feel in what space we find ourselves" (Böhme, 2006, p. 89).

But this genealogy of local space needs to take one step further. As we have seen, the aqualitative-abstract local space, rarely inclusive, presupposes the directional space, whose kinesthetic and synaesthetic aspects make, for example, a certain dwelling more inclusive than another. However, directional space has its condition of possibility only in the original space of the vastness, totally extraneous to surfaces and physical-geometric dimensions: this is the already mentioned (see Chapters 6 and 8) "here" of the primitive presence/present. The latter can appear in an inarticulate and unintentional form, as in the case of climatic or thermal, kinetic-posterior or anterior optical-kinetic space (the latter are the only ones that allow for our movements and for a dancer's exceptionally fluid confidence), or in the articulated and perceived form recognizable in spatial phobias and even in ecstatic-orgasmic experiences. This original space (of vastness) exhibits a voluminousness that, acting as an authentic apriori – not of pure intuition as Kant claimed, but of our felt-bodily feeling – makes us feel included (or not) in the deepest sense of the term.

The adult experience, integrating these three forms of spatiality, exemplarily in dwelling and in architectural space – understood as the "extended present" (Nowotny, 1995,[2] p. 87 ff.) that acts as an existential-aesthetic envelope indispensable to our routine – can only experience inclusiveness as an affective state derived from the constant interweaving of the three spaces. In fact, inclusiveness can be found in the "ecstatic" qualities atmospherically irradiated by individual points of the local space, dynamically tonalizing their surroundings, as they welcome or reject us. Inclusiveness, above all, can be found in the affordances of the directional-vectorial space (centrifugal, centripetal, omnilateral, undecided, etc.) as well as those, all the more intense and rare, of vastness (satisfaction, despair, etc.). Now, having established that all three spaces are atmospherogenic, albeit to a different extent and, moreover, almost always in their interweaving, it is necessary to clarify what I mean by "inclusive".

It is quite likely that one would move differently, for example, in a Nordic and in a Mediterranean landscape (Dürckheim, 2005, p. 49). This is true *a fortiori* in a differently built and furnished space, which therefore expresses its specific gestural affordance (in a broad sense) in the anticipation of the user's first ingressive and then ambulatory-internal movement. This approach obviously presupposes that one thinks of the architectural space as a *Gesamtkunstwerk* irreducible either to its glossy journalistic photogenicity or to its reductive digital output, which by definition miss essential variables such as climate, atmosphere, physicality and scale (Pérez-Gómez, 2016, p. 148). Being a stage for sensual atmospheres (cf. Wigley, 1998, p. 20) rather than a form of intelligent problem solving, architecture actually finds more than a "niche" (as Mallgrave, 2013 reductively claims, p. 88) in the generation of (atmospheric) feelings. Of course, its atmospherogenic charge should be aimed not so much at facades and Potemkin villages – in short, a banal orchestration of effects (cf. Leatherbarrow, 2015) – but, on the contrary, at enabling inclusion or exclusion thanks to different means such as things and colours, shapes and sounds. Indeed, as already mentioned (see Chapter 9), sometimes it is the characteristic sound of a building that radiates invitation or rejection, hospitality or hostility.[3]

Then, inclusive buildings are especially those that, with their splendid "resonance" and their (even psychic) temperature, tell me that I am safe and that I am not alone (Zumthor, 2006, pp. 33, 35), thus differentiating themselves from the bureaucratic coldness of excluding ones. This is true, however, only if we see architectural objects as "orientations, kinetic suggestions, marks" (Böhme, 2006, p. 113), in short, the "frozen" gestures of the perceiver's anticipated kinetic reactions. These skillfully staged spaces (remember Boullée's talking architecture but also Kiesler's narrativity, cf. Pérez-Gómez, 2016, p. 104 ff.) invite percipients to immerse or distance themselves, pathically modulating – more soundly than a museum installation – the pericorporeal space of those who observe and, above all, walk through and inhabit them. But this presupposes that the semiotic-communicative (sender-message-receiver) paradigm and the interpretative-hermeneutical paradigm be replaced by a performative model (cf. Jäkel, 2013, p. 32 and Pérez-Gómez, 2016, p. 146), centred not on what a space "represents", but on what is "felt" in it. It presupposes that the centre of reflection be embodiment, that is, a co-presence of human and situations understood as sufficiently "porous" poles. In many languages, the trivial question "how are you?" coexists with the question "how do you find yourself" ("come ti trovi" in Italian, "como te encuentras" in Spanish, etc.) thus integrating the topological sense with the existential sense and thematizing the already mentioned felt-bodily communication between percipient and world – symmetric or asymmetric, depending on whether the second pole is organic or inorganic – as finally emancipated from Western ocularcentrism. Far from simply being the creation of vast sculptures (Böhme, 2006, p. 175) which, like photographic images, are subject to a frontal-distal observation that is almost perfectly describable in the third person, architecture presents itself here as a synaesthesic lived experience, especially today, within the framework of a growing re-sensualization (Pallasmaa, 2005, p. 37).

By becoming aesthetized and thus also qualifying its space from a motor point of view, every architectural form is able to generate an inclusiveness (as "effect" or as "imageability", cf. respectively Sitte, 2006 and Lynch, 1960) that "tonalizes" the entire subsequent experiential flow. There are several fanciful bio-architectural theories, according to which our biological system would be ancestrally predisposed to feel at home in the savannah(!), as opposed to, say, a collection of glass buildings overlooking a parking lot without vegetation (Mallgrave, 2013, p. 102). Now, regardless of such ideas, we could hypothesize that an inclusive atmosphere (as mentioned) guarantees a spontaneous and hermeneutically circular interlacement between the repertoire of architectural gestures and that of the user's spatial and felt-bodily experiences. The area of the lifeworld that is thus kinesthetically and synaesthetically lived becomes a real atmospheric isle. Architectural gestures generate inclusive atmospheric isles when their elements (wall, window, door, staircase, niche) suggest (obvious or virtual) motor figures in the percipient, inviting them to linger in that space. Instead, other motor figures, such as an angular building (think, prototypically, of New York's Flatiron skyscraper), suggest inclusiveness for those inside while excluding altogether those who are outside. The unusual angularity and insulating projection of the building, in particular if perceived on its own, arouses the impression of an excluding power aimed at social discipline; if perceived in a different context (climatic, perspective, urban), it expresses instead a certain eccentricity, and yet certainly not inclusiveness, as in any case it (literally) "puts us in the corner" (cf. Hasse, 2012, pp. 101–120). The least that can be said, at this point, is that inclusiveness is a gesture that intertwines the three types of space and generates a certain immersiveness. But do we really know what "immersive" means?

Driven by the iconophilic apology of both the performativity of contemporary art and the alleged pervasiveness of the virtual, it is all too easy to say that we live in the age of immersiveness, meaning an illusionary mediation that appears immediate and therefore able to promote an intensified vitality. After all, cultures have always been "collective conditions of immersion in air and sign systems" (Sloterdijk, 2012, p. 84), and there is no need to find proof of this, say, in the Las Vegas mimetic reconstruction of piazza San Marco. The fact is that, as per usual, even "immersive" is an ambiguous notion. In its physical sense, only the ordinary perception of felt-bodily space – by virtue of which we are always "outside", near things, immersed in them – can be said to be immersive. Instead, in the figurative (emotional) sense, one can define immersive any hyperrealistic strategy that makes the absent present, momentarily annulling with technological artifices the distance between medium and receiver, but still limiting itself to functionally imitating ordinary experience through a hybridization of felt-bodily space and representational space (Böhme, 2013 – the nausea produced by flight simulators is real, just as undergoing telesurgery has real effects on one's body). Such immersiveness (in a metaphorical sense) is not so different from what is experienced by a reader captured by a novel, by a spectator ready to identify with a film's characters, by the observer of a canvas, and even by a believer anticipating happy eschatological worlds or by a purchaser absorbed by the consumerist game.

Today's over-rated immersiveness is therefore nothing but a playful "spectatorial" play with a felt-bodily and fully pathic dimension, which, if anything, offers "relief", given that a truly immersive image, i.e. one that is not "only visible", would *ipso facto* cease to be an image. Therefore, having reiterated that an authentic immersion only belongs to ordinary experience (cf. Wiesing, 2014, p. 131 ff.), that is, to the situation in which we concretely and individually participate in what we perceive, which in turn could in principle perceive us ("ageing" with us), we can well say that every densely atmospheric space that implies the real and felt-bodily presence of the percipient, and their full involvement, is also an immersive space. A space that, involving us at first, then makes us feel excluded could be defined immersive only in a broad sense. Keeping this in mind, and not abusing this qualification, it is certainly legitimate to consider the inclusive space a variant of the more general and generic immersive space: more precisely, it is the variant that, as we shall see now, necessarily involves intimacy and familiarity. Therefore inclusiveness, in the proper sense, is a specific architectural-gestural quality. In fact, a church must induce devotion, a concert hall the will to listen, a court fearful respect for the law. Evidently aimed in these cases not at inclusion but at stimulation, and perhaps for this reason led by projects that are more sympathetic than parasympathetic (Mallgrave, 2013, p. 112), space is certainly relaxing-inclusive when it is habitable (in the full sense). Because of its atmospheric and quasi-thingly resonance, a space is inclusive, as well as oppressive, liberating, etc., always in terms of an exteroception mediated by a non-strictly organic-anatomical proprioception and enteroception. That is, when its affordances suggest, as already mentioned, the (real or virtual) felt-bodily behaviour of dwelling in it. If it contains a table and an armchair, for example, (as opposed to a table alone) a certain environment leads me to stop and sit. If there is only one chair, placed on the longer side of a table, I would even be led to stop to work and read, whereas many chairs in front of a table suggest that I should stop only if I intend to attend the public event to be held there. A ladder directs me from bottom to top, of course, but it is "how" it does so that constitutes the inclusive (or not) atmospheric gesture that interests us here.

We started from the lived experience of a building (Chapter 9) and, passing through the atmosphere of the home (Chapter 10), we arrived to dwelling as a lived space, finally wandering inside it to better understand what its inclusiveness could consist of. We rightly feel a bit tired and, rather than lying down (full passivity) or remaining standing (full attentional operation), we choose to relax on the armchair.[4] With its backrest and its armrests, it comfortably wraps around our body, almost acting as an archetypal enclosure (and therefore as a dwelling), and with its position it avoids the promiscuity of glances, thus ensuring a "relaxed social interaction which makes no demands, which is open-ended but above all open to play" (Baudrillard, 2002, p. 45). And then, perhaps, we also close our eyes, shrinking away, through a sort of "inside look", from the local, isotropic and therefore faded space. Thus we access an authentic "new world", characterized by a vastness that only relative space normally conceals, and which instead is marked by orientations closely linked to the number of feelings that visit us, to the concert of "memories, needs, temptations, solicitations, delicate atmospheres, felt-bodily impulses" (Schmitz, 1977, p. 207; cf. also 2007, pp. 75–76). The atmospheric potentiality of dwelling seems decidedly at its peak here, since spatiality is transformed into *spaciousness*, that is, into a vital dimension that is so extra-dimensional as to be incomparable (Minkowski, 1954 insists on this), where, in fact, the ease of living is transmitted to anyone who shares the same atmosphere. This can be seen, for example, by the fact that sometimes people feel good together without having to say a word (Soentgen, 1998, p. 81), or that, casting our gaze on a partially hostile exterior, we can experience and savour a controlled and controllable shiver.

In the atmosphere of feeling at home, an important role is also played by some specific kinetic possibilities, which amount to having an absolute spaciousness founded not on metric geometry but on "measuring distances in terms of 'our hands and legs' and having the main directions loaded with the bodily intentionality that makes a geometrically measurable space a familiar domain" (Giordano, 1997, p. 18). That's why the steps taken at home are so qualitatively specific that they cannot at all be added to those that taken outside. This is why, moreover, only an inclusive dwelling guarantees genuinely reversible directions, which are rather rare and in any case dysfunctional outside: in fact, only at home does one move without particular justifications from one room to another, using the same couch to sit or to lie down, perceptively moving in non-predetermined directions without excessively demanding torsions (functional, psychological but also bodily). Ultimately, at home, one can "lose oneself" in *lebensweltlich* situations and reassuring habits, controlling one's own time without being either socially sanctioned or condemned to the marginal and aestheticizing figure of the *flâneur*.

For centuries, in Germany, this culture-cultivation of atmospheric feelings effused in the interior space – this "art of staying at home", fulfilling a need for both human contact and privacy, which gradually turns something uncanny into one's home – has had a name that, while being in itself semantically vague and often declared untranslatable in other languages,[5] is notoriously akin to petty-bourgeois pharisaism: *Gemütlichkeit*. For many, this term evokes a merely comforting surrogate and certifies the typically German (cf. Mitscherlich, 1965, p. 129 ff.) psychic palpitation that often, since the Biedermeier, confuses home fetishism and the lived experience of dwelling. But is it really like that? Is the *Gemütlichkeit* really only the trivial compulsion to exorcise the external asymmetry through internal symmetry, to saturate the house with objects, to avoid the threatening *horror vacui* outside? Or, with an equally kitsch choice,[6] to rarefy objects, in order to oppose the intolerable fullness of the external world? Does it really evoke only the stale set of family interests, the live embalming in relation to which one should instead state that "the house is past" and that "dwelling, in the proper sense, is now impossible" (Adorno, 2005, p. 38)? Does it really only suggest the image of dressing gowns and slippers, of good living rooms, perhaps of geometric decorations, whose relaxing effects are due to the fact that "the beholder is involuntarily affected by the slow, calm mode of production" (Stritzler-Levine, 1996, quot. in Sparke, 2008, p. 99), that is to say, a whole series of largely trivial choices, inevitably colonized by marketing?

Let's have a better look, then, devoid of eschatological-political blinders. The atmosphere of *Gemütlichkeit*, first of all, is always discreet and sober, so fragile as to be threatened by any unusual or unexpected noise or event. But its fragility, by which it becomes something less obvious only when it feels threatened, does not transform the habit of "legitimate selfishness" (Schmidt-

Lauber, 2003, p. 66) it produces into a choice that is necessarily contrary to good taste. A comfortable intimacy is in fact the basis of much of our aesthetic taste. It spiritualizes and "authenticates" on the pathic-axiological level – that is, it individualizes in the proper sense – things and places that are otherwise relatively indifferent (from furnishings to clothes), by virtue of the subjectification and ritualization of a daily routine based on leisure time, on relaxation and, why not, on sociability – even if circumscribed, perhaps in the name of "distinction", to a few select people. *This* intimacy is certainly founded on small talk, or at least on discourses devoid of controversial edges; on various forms of cuddles (with people and/or pets); on moderate heat and soft light, able to bring out the depth that full sunlight or artificial light normally put in the background for the benefit of "cruel" close-ups;[7] and on the "moral theatricality of the old furniture" and on the ticking clock, as "the clock's sound assimilates the place to the inside of our own body" (Baudrillard, 2002, pp. 18, 24). More generally, it is based on a *soundscape* that does not prevent concentration (be it for reading, prayer or dialogue).

But any codification of this atmospheric condition – as we know all too well – risks being ridiculous or at least marked by affectation. In fact, in order to generate it, it is not enough to have a sofa[8] and a fireplace, curtains and carpets, decorative plants and ambient music, cushions and wooden floors, elegant wine glasses and/or "exquisite" coffee cups: it is also necessary to blindly give in to otherwise threatening feelings – something that often, for us, coincide with wellness. This specific living atmosphere, in its most superficial traits, is undoubtedly due to the modern psychologization-moralization of the affective, and is not entirely observable in the third person, as it is even threatened by reflexive investigation,[9] even if only verbal. This should not be too surprising, as it is one of the absolutely subjective facts that are too often sacrificed in favour of so-called objective ones. It is one of the necessarily transitory vital experiences that, although made possible by an authentic felt-bodily liturgy, cannot be *entirely* intentionally produced. In fact, one might try as hard as one wants to stage atmospheric (*gemütlich*) situations by, say, resorting to a bath, a soft pillow, and so on: however, the atmospheric nature of dwelling, like the one that generally depends on design, is rarely perfectly plannable, since its spontaneous and involuntary character is clearly an indispensable component of the effect that it arouses.

Stigmatized by any stylistic-political avant-gardism as regressive, late-Romantic, provincial, petty-bourgeois and *ipso facto* reality-denying (because reality would be tragic in principle!), the atmosphere of *Gemütlichkeit* proves to be, in my opinion, completely unavoidable: and not only since when bourgeois society, precisely with its cult of interiors, tried to "differentiate itself" both from the aristocracy and from the working class (Schmidt-Lauber, 2003, p. 155). In fact, every time that the architectural project conforms to the "domestic ritual, translating it into distances, passages, sequences", it becomes a "map of a domestic universe revolving around fixed orbits and composed of 'things'" (Vitta, 2008, p. 229), and this necessarily generates the comforting protection-enclosure without which our emotional life would be greatly impoverished. As we know, without a comfortable enclosed space in which to cultivate our feelings, our affective life would be very meagre indeed, lacking the embodied scaffolding that, alone, allows our emotions to emerge or at least to become more precise (to use the terms of my atmospheric enactivism). Of course, the (homely) world of life once revolved around the almost exclusively "horizontal" (sleep, sickness, eros) and extremely ritualized axis of the bedroom. Then, after World War II, it was, so to speak, transferred to the hall with its "frontal" representative function,[10] and then in the living room with its TV totem. Today, perhaps also because of the media flood of "Masterchef" *et similia*, it is in the kitchen, which has become a veritable "corporate throne room" whose power does not seem (for now?) usurped by what the obsession with fitness defines "post-bath" (Vitta 2008, p. 239, p. 251).

Well, whatever its focus (in the light of proxemic values that are clearly relatively changeable), and whether it aims at a quiet breakfast, a relaxing bath or a pleasant reading, dwelling is and remains authentic only when it coincides with a situation that is (anyway) *gemütlich*. In other words, a situation that, as seen in every atmosphere, "supervenes" on indispensable objective catalysts (Schmitz, 1977, p. 264 ff.), like (mild) temperature, acoustic insulation, chiaroscuro lighting, and maybe even the architectural solutions (in the broader sense) which, unlike the artificial ones that are not really "lived", find no harm in aging (Fromm, 2008, p. 71). In fact, in exceptional cases, the dwelling atmosphere may certainly depend more on the subjective link with the place than on specific material components (how else to explain that house stairs are sometimes more disturbing than the road?). However, this atmosphere normally supervenes on, and also depends on, such necessary (but not sufficient) conditions. Indeed, it is only released in the presence of expressive qualities that are strictly immanent in a certain environmental setting[11] – provided, of course, that both the perceiver's felt-bodily disposition and the situational context are "normal". In fact, in the case of obvious and contingent divergence, for example, between excessive housing comfort and a serious external socio-political situation, or of tension due to a guest's awkward remark, even the best-planned atmosphere could fail, or at least be felt as inappropriate.

The aforementioned distinction between prototypical, derivative and spurious atmospheres can be useful here too, in order to distinguish between a) prototypical inclusiveness, supervenient on profound structural modifications of interior architecture (just think of the enormous difference between an open space and an apartment whose rooms branch out of a long corridor); b) a derivative one, based on a residential satisfaction following from the interaction between subject and object; and finally c) a spurious one, which appears inviting and welcoming even for completely idiosyncratic reasons, such as when someone feels welcomed and protected also by the most anonymous and public space in the crowded city streets. Whatever one's different individual topological sensitivity, I do not think it can be said, in a totally relativistic and ultimately projectivistic way, that "there is not a space in the world which cannot be intimate for somebody" (Wydra, 2014, p. 68), or that "the 'uncanny' is not a property of the space itself nor can it be provoked by any particular spatial conformation", as there are architectures "that, from time to time and for different purposes, are invested with uncanny qualities" (Vidler, 1992, pp. 11–12). This is because there are phenomenally attestable invariants that prevent us from perceiving inclusiveness, for example, in buildings with sloping walls and floors (as in the park of Bomarzo!), in areas subject to impetuous gusts of wind, in high-traffic roads, around columns that do not seem comfortable to lean on (Mallgrave, 2013, p. 138), or in houses haunted by "insufferable gloom", as is said, emblematically, of the aforementioned House of Usher (Poe, 1856, p. 291; cf. Griffero, 2018). No subjective and projective mood is in fact able to transform these spaces into places whose atmosphere is fully comfortable, hospitable and inclusive.

We have seen that, like every enclosure (up to the extreme case of the armchair), dwelling breaks universal continuity. But it is an intrinsically porous enclosure, since it has areas that are obviously critical and ambivalent (threshold, roof, windows, wall, etc.), as they put in contact inside and outside, and whose overcoming always generates – and not only in children – a peculiar "anguish of the border" (Schmitz, 1977, pp. 229–232, 241). Like dwelling, its atmosphere therefore shows ever-renegotiable limits, destined to be continuously crossed. It is precisely because from time to time it is trespassed and reconstituted that the border generates, in the final analysis, the qualitative differentiation typical of the dwelling atmosphere. As seen, the latter is the cause of both the immunizing and apotropaic management of external sentimental powers as well as of their control (through a filtering that does not completely anaesthetize them) aimed at preventing, once tamed, their external "bleeding".[12] In short: we only feel at home if at other times, and maybe during the same day, we have felt *not* at home. And this analogy with the sacred enclosure raises four unanswered questions.

a) Is the most intimate, and therefore more "sacred" enclosure of dwelling the place of maximum familiarity and maximum comfort? Is it the more evocative the more inclement the external (climatic) conditions?[13] Or is it rather, as in the case of the church as the home of the divine, the epiphany of something totally different, more mysterious and disturbing than the external and profane space?

b) Also, while maintaining the neo-phenomenological difference between dwelling intimacy and psychic introjection (Schmitz, 1977, p. 213), perhaps the insistence on the enclosure, *a fortiori* if purely symbolic and not material, is relatively incoherent with the radical atmospheric and neo-phenomenological externalism. It is no coincidence that Hegel, far from refractory to the introjective principle, underlines the primary role of Christianity in performing this delimitation-introjection.

> Just as the Christian spirit concentrates itself in the inner life, so the building becomes the place shut in on every side for the assembly of the Christian congregation and the collection of its thoughts. The spatial enclosure corresponds to the concentration of mind within, and results from it [...] The impression, therefore, which art now has to produce is, on the one hand, in distinction from the cheerful openness of the Greek temple, the impression of this tranquillity of the heart which, released from the external world of nature and from the mundane in general, is shut in upon itself, and, on the other hand, the impression of a majestic sublimity which aspires beyond and outsoars mathematical limitation [...] The space inside therefore becomes totally enclosed whereas the Greek temples often had open cells in addition to open colonnades and halls.
>
> (Hegel, 1975, II, pp. 685–687)

Against this approach,[14] one should first of all note that even the most interiorizing church still needs to be built in a "true", i.e. numinous, space. Moreover, sanctuaries are neither edified nor chosen at will, but can only be found thanks to the art of orientation.[15] Which means that, even in the case of dwelling, the interior remains bound to a specific external atmosphericness, and that dwelling *qua* feeling at home always presupposes the contrast with the equally lived and felt-bodily condition of *not* being or feeling at home (Vitta, 2008, p. 299; Bürklin, 2008, p. 160). In other words,[16] it presupposes that the threshold is regularly crossed and that therefore the care of the interior, in its qualified extroversion, only excludes the disturbing external vastness and not the duly filtered one (Schmitz, 1977, pp. 268–269).

c) One can also wonder which is more conducive to the cultivation of atmospheric feelings: a space cleverly designed to manipulate the mood of those who dwell in it, or a space that simply suggests its atmospheric potential, without trying to control the unpredictability of its effects? Consider a house prepared with maniacal care for comfort and social appearance, in which every detail, almost as a discrete planned stimulus, tries to mechanically generate a certain *Gemütlichkeit*: it would probably be irritating. It would be much more pleasant to be in a house whose atmosphericness, far from being a thetic-intentional pole, is an uncertain and perhaps even only transitory condition of possibility. In this sense, one could also admit that, in dwelling, planning can only concern the condition of possibility of the actual manifestation of atmospheres – affective situations that are so extra-psychic that they cannot be intentionally mastered.

d) Finally, one might say that this pathic-atmospherological approach to dwelling sacrifices any political implication. This is far from true, if by "political" one refers not so much to the critical or party discourse on society, but rather to the "disposition of feelings sedimented in the lived bodies of individuals" (Hasse, 2009, p. 224). In other words, there are indeed political implications to this approach, if "political" refers to how an atmosphere imposes its authority (cf. Griffero, 2017a, pp. 29–38), generating (in the case now investigated) a compromise solution between the resident's freedom of intervention and the forms of power inscribed even in the most neutral housing situations.

The important thing, however, as in the already examined case of the urban atmosphere, is not to throw the baby out with the bathwater. In other words, not to identify the inescapable need for an atmosphere of protective intimacy with a superficial apology of petty-bourgeois private comfort. I do not believe, as Anders does, that familiarity is always a pseudo-familiarity – that is, an effect of alienation. It is true that no one can completely master atmospheres (of dwelling or otherwise). However, the solution does not lie in rejecting present dwelling for a new hypothetical dwelling suggested by new needs and by an unspecified new social order (Adorno, 2005, p. 39). In a much less utopian sense, I am content with being more than a mere consumer of living quarters, escaping the ever-present fetishistic risk of an aesthetics of the sofa[17] – or even of garden gnomes – through a capillary phenomenological analysis of the elements and their expressive qualities on which a balanced living atmosphere supervenes. If the alternative to *Gemütlichkeit* dwelling is nomadism, that sort of global hotel that promotes itself by tricking us into thinking that we will feel at home everywhere in the world, well, this alternative does not seem preferable to me, above all because, far from allowing us to enjoy the atmospheres of the world in a filtered form, it gives them *carte blanche*,[18] or, as often happens, debases them up to cancelling them. Ultimately, Balzac could be right in saying that the "privilege of being everywhere at home is the prerogative of kings, courtesans, and thieves".

Notes

1 Starting, at least, from Gallese (2005). For an application to architecture cf. Mallgrave (2013, p. 133 ff.). I share the latter's theses on architectural affordability (and even on atmospheres, in particular that of light: ibid., pp. 157–161), but I am a bit wary of his excessive confidence in the neuroscientific approach.

2 On "ecstasies" (as opposed to "properties") cf. Böhme (2017b, pp. 37–54).

3 Cf. Pallasmaa (2005, p. 50), who even in his other works wishes for the petrified silence of architecture against communication abuses.

4 Here I follow the example used by Schmitz (1977, p. 207 ff.). But, with healthy impertinence, Soentgen (1998, p. 79) suggests, remembering Tanizaki (1977, pp. 3–6), that the best example of relaxation could then be the toilet!

5 Here one could use the English terms "lounging", "cocooning" (cf. Schmidt-Lauber 2003, p. 12).

6 In fact, "even the most rigorous minimalism can fall into kitsch" (Vitta 2008, p. 317).

7 For the atmospheric potential of dimmed light cf. Griffero (2017a, pp. 103–112).

8 A true "communicative piece of furniture", according to Warnke (1979, p. 677).

9 As Schmidt-Lauber (2003, p. 19) acknowledges immediately.

10 Baudrillard (2002, pp. 46–47) speaks of furniture based no longer on the "solemn backdrop for affection" but on "an equally ritualistic décor of reception".

11 I believe Schmidt-Lauber (2003, p. 209) agrees on that.

12 As many historical sources show, the sacred enclosure was also aimed at preventing local spatial deities from abandoning their residence and wandering outside their original dominion.

13 Cf. Baudelaire (1996, p. 114), for whom winter adds to the poetry of a house. It needs "as much snow, hail, and frost as the sky could possibly deliver. [Those who live there] must have a Canadian winter or a Russian one".

14 According to Schmitz (1977, pp. 283–286), Hegel here fails to note that Greek temples were also built within a sacred *temenos*.

15 As noted by van der Leeuw et al. (1955, p. 398).

16 Think about how today's huge windows make the exterior a sort of indispensable "media carpet" (Hasse, 2008, p. 122), if not an explicit aggressor of the intimate or private realm (Baudrillard, 2002, p. 43).

17 Often connected to the narrowness syndrome, perhaps typically German also because it originated from the historical territorial fragmentation of Germany (Bausinger, 1988, p. 49).

18 For Schmitz (1977, p. 245) the absence of true boundaries gives way to the affective uncanny.

12 Found or created?

The *genius loci* as an (atmospheric) experience

Alexander Pope once said: "Consult the Genius of the Place in all". However, the concept of *genius loci* is both theoretically inexplicable and practically (existentially) certain. What Augustine says of time also holds for the genius loci: in fact, "If no one asks me, I know; If I want to explain it to an interrogator, I do not know", and the tendency is rather to make a metaphorical use of it. Yet even those who, in the spirit of the Enlightenment, mock any belief in spirits, stigmatizing it as an anachronistic animism, would not completely exclude the possibility that certain places exert a special "spiritual" influence on people (whatever that may mean) by virtue of their specific "character", which precisely for this reason is considered worthy of protection. Even the coldest rationalist, in short, "feels" that there are certain places whose identity spirit should be preserved from the threat of *placelessness* (cf. Relph, 1976, p. 6).

The pier of Ostia (Figure 12.1) may appear as "visual chaos" (uncertain weather, disorderly seaside architecture, forgotten road signs, narrow perspective because of the pier wall, etc.) devoid of any *genius*. But things change as soon as we frame the same place from a slightly different topological (even just ocular) perspective. In fact, it is enough to look at the same pier by moving just a few centimetres to the left (Figure 12.2), albeit at the same hour and in the same weather conditions, for the atmosphere to change a little (more centred perspective that shows, in the long shot, buildings and street lamps placed neatly and symmetrically, etc; for a similar example cf. Führ, 1998). Besides, why couldn't visual chaos be the specific spirit of a disorderly Roman beach?

This example already shows the theoretical unsatisfactoriness of both scientific and popular discourses on the *genius loci*, given their metaphorical vagueness. In fact, the argument adopted in favour of place-identity is usually the cliché that the *genius loci* is the *quid* that transforms an anonymous space into an intensely expressive lived place. As plausible and unprovable as any other cliché, this argument expresses the inescapable need for some correspondence between internal landscape and external landscape. This need has become even stronger with the end of the "asceticism by exclusion" of the International Style, as well as with the proposal, suggested (cf. Casey, 1993,

Figure 12.1 The pier of Ostia version A

p. 110, Tuan, 1977, p. 3; and Malpas, 2004, p. 31 f.) by the spatial turn (cf. Günzel, 2014, pp. 25–43) of many humanities, of seeing the condition of possibility of the isotropic-naturalistic space in the inclusion in a lived-qualitative space (cf. Griffero, 2014b) that serves as a meaningful location (Cresswell, 2004, p. 7) and acts as a place (in the full sense of the term).

The literature on the *genius loci*, unfortunately, often indulges in metaphorical and rhetorical expedients, relying on the self-evidence of terms such as character, quality, livability, taste, feeling, environment, essence, resonance, presence, aura, harmony, grace, charm, *decorum*, etc. (Hiss, 1990, p. 15). But the need to defend the identity of places from topological globalization, from modern cartographic reductionism[1] and from banal touristic *vedutismo* (as a merely distal and ocularcentric experience), does not seem able, in this way, to contrast the post-Enlightenment eliminativism. The latter can be summed up (cf. Kozljanič, 2004, II, pp. 57–69) as the degradation of the *genius loci* to a

Figure 12.2 The pier of Ostia version B

residual product of a mythical-archaic (childish, prelogical, anthropomorphic, evemeristic) mindset and of space to an a-semantic screen which is the object of arbitrary projection-constructions. And it is not enough to adopt the evolutionary argument according to which a *genius loci* (and an intimately aesthetic value) would be recognizable in any landscape that can somewhat be traced back to the savannah as the original human habitat, not clarifying why this *genius* has often been attributed to places that instead have very little to do with the savannah (high peaks, deserts, islands, etc.).

Not only is the notion of *genius loci* metaphorical and vague, but also abstract and secularized. Indeed, in Greece the *genius loci* used to be understood as a *daimon* (especially nymphs), which was responsible for ecstatic raptures and a pneumatic-creative energy often experienced in caves, mountains, woods and even houses;[2] then in Rome it lost its numinousness and its original valence of chthonic and disturbing semidivinity, to the point that it was seen as the protector of every person and every place. The poetic or, worse still, commercial/real-estate trick of asserting the special spirit of a certain area for cultural *and* natural reasons (as if this conjunction was not problematic), is but the remote effect of this Roman inflationism, which was so immanent that it saw the *genius loci* as sometimes entirely personal and sometimes so collective as to influence a whole people, a race, its places, and so on. This move, indeed, also risks undermining even the architectural choice of using the spirit as a pre-planning intuition indispensable for the organic-ecological and hierarchical-functional overcoming of an otherwise simply geometrical project (Neddens, 1986, p. 24 ff., pp. 91–92). We are not told, in fact, how to identify (and understand) this spirit, or what its added value is compared to an urbanist who does not pay any attention to it.[3]

An equally traditional move consists in the illusion that semantic clarification passes through the proliferation of (physical and spiritual) meanings of the term. Is it really useful, in order to generate/preserve the spirit of a place, to identify 253 different topological patterns ("high places", "sacred sites", "access to water", "accessible green", etc; cf. Alexander, Ishikawa and Silverstein, 1977)? Is it useful to state that the *genius loci* necessarily requires several elements, like architectural style, climate, natural setting, memorial and metaphorical value for the residents, material used on the basis of specific craftsmanship skills, constructive sensitivity and historical-cultural specificity, popular values lingering in it, high quality level of the visible and accessible environments, daily urban and seasonal activities that take place there, meanings and symbols conveyed by it, and so on (cf. Garnham, 1985 and Steele, 1981, p. 99)? Or is it enough to define the *genius loci* as an almost personal-identity animic entity, whose potentialities come to light thanks to anthropic intervention (cf. Müller, 2009, pp. 126–132)? Or perhaps it should be defined as a sphere endowed with its own peculiar (even geomantic) dynamic,[4] or as a spatial unit that is accessed by crossing a threshold (even symbolically). Or else as the mythical or historical mnestic trace of a collective identity, or as a space full of accessible information and effective factors … In identifying both its presumed material components and its presumed symbolic-cultural values, too often have thinkers fallen into inflationism which, while rightly understanding the coalescence of the elements constituting a "situation" (cf. Thibaud, 2015, pp. 229–230), naively believes that proliferation may compensate for a probably unavoidable theoretical vagueness.

These theoretical difficulties also affect the most serious studies on the subject. In his famous phenomenology of places, Norberg-Schulz, in fact, conceives architecture as a concretization of existential space and as a pregnant visualization of the *genius loci*, while obscurely defining (cf. Shirazi, 2008, pp. 1–3) this *genius* as something that should be taken into account (something to dialogue with) by anyone who wants to live in an authentic way (cf. Norberg-Schulz, 1980, pp. 5, 7–8, 10–11, 15–16, 170, 179) and "feel at home in it" (although it may be surprising and can be variously interpreted). Almost nothing is said about the extent to which the spirit of the place, in its "varying degree of invariance" (ibid., p. 18) – in line with the rhetorical-aesthethological topos of "variations on the theme" – resists transformations and reinterpretations over time, if and when these are misrepresentations or adaptations, how to identify the theme, or why some places are so non-atmospheric (ibid., pp. 21–22, 72) as to be devoid of a *genius* (as if there couldn't be a negative *genius loci*!). Norberg-Schulz's phenomenology is also undermined by being fully "external", i.e. based only on aerial views, wide shots and above all on the facade of buildings, in short on a scenographic and ocularcentric frontality[5] that ends up underestimating both the *genius* of non-urban places and the perceptual role of the body in motion (Shirazi, 2008, pp. 6–10), especially of the felt body (*Leib*) in a (neo) phenomenological sense (cf. Griffero, 2017b).

Even his examples appear unconvincing. The physiognomic qualities attributed to Prague, Khartoum and Rome, in fact, do not appear to be sufficiently specific, nor actually invariant with respect to the physical and historical transformations of those cities, only appearing salient if, with a fatally circular reasoning, one has already preliminarily chosen them as the parameters of the analysis. For example, consider the feeling of being "inside" even while being outdoors ("idyllic urban interior"): isn't this the spirit of any Mediterranean city rather than just that of Rome?[6] The Italian capital, and even Piazza Navona,[7] are attributed a spirit consisting of an olympic humanization (Colli Albani) of nature (of the archaic "chthonic forces" and forres of Etruria) and capable of gathering "all the basic categories of existential meanings" (Norberg-Schulz, 1980, p. 164): well, doesn't this depend too much on cognitive elements (geography, history) that are sometimes even sensorially concealed (archaeological finds) to be still valid as perceptive-phenomenological? Finally, aspecificity and partiality also characterize Norberg-Schulz's classification of the *genius loci* in ideal-typical forms:[8] for example, the Nordic-Romantic landscape and architecture are founded on the search for a protective shelter; the humanist-classical ones are dominated by the union of man (inside) and nature (outside); the cosmic-desert ones are destined to monotheism in their abstraction. All of these, however, are irreducible to the physiognomic character hypothesized here (as this character is supposed to be present in varied landscapes). Above all, Norberg-Schulz's classification disregards the *genius* of "mixed" or more complex places and of places like jungles, tropical islands (cf. Kozljanič, 2004, II, pp. 314, 320) etc., which appear salient for non-European scholars, interested, like for example Watsuji (1961), in significantly different types of climates (monsoon, desertic and pasture climate).

The theoretical dissatisfaction that one experiences while reading Norberg-Schulz's (high-level) observations also becomes practical if one tries to understand how to implement a *genius loci* or even design it from scratch. When trying to identify the *genius loci*, one cannot ignore the topography of a place, its physical-climatic – and therefore also qualitative-affective – discontinuities, the characteristics of those who inhabit it (humans and animals), its soundscape and smellscape, the means of transport that cross it, etc.[9] And yet the concrete imageability that should characterize a place, be it a *locus amoenus* or *terribilis*, presupposes that we pathically and felt-bodily expose ourselves to it, establishing a sort of "place ballet" (Seamon, 2013, p. 13), and perhaps counting, as Klages would say, on a mood that is receptive because it is not too "robust"[10] – in other words, a mood able to feel and be impregnated by an affective tonality that is as impalpable as the atmosphere.[11] The latter, in my hypothesis, is exactly what makes the environment non-neutral and finds a precise sounding board in our felt body thanks to a pathic perception[12] (with inevitable existentialist consequences).

But the first step for an atmospheric approach to the spirit of places implies getting rid of the idea, controversially attributed by Christian authors (Prudentius, *Sym.* II, 446–449) to the polytheistic ones, that "nullus enim locus sine genio" (Servius). The *genius* does not reside anywhere, nor can it be transported or interchanged at will.[13] Indeed, as noted by Ovid (*Met.* 3, 157–162), it is *ingenium naturae*, not *hominis* (Kozljanič, 2004, I, p. 76), as it mainly signals the uncanny-numinous nature – thick woods, solitary places and dense shadows, which are not at all interchangeable. For both Seneca and Rudolf Otto, these places create the sacred (or numinous, cf. Harrison, 1903). This uncanny-numinous nature is perceived neither by those who habitually live in a certain place satisfying their needs, nor by those who simply complain for its qualitative impoverishment,[14] but especially by those who are pathically and felt-bodily raptured by it. In the form of an affective-expressive leitmotiv (Kozljanič, 2010–2011, p. 160), whose condensation points are always only relatively autonomous, the spirit of a place, just like the prototypical atmosphere, is embodied in a certain space before being historically implemented or even idiosyncratically projected in it by culture.[15]

This ranking of the different spirits of places, for which places can be more or less atmospheric (i.e. more or less able to express a *genius*), is obviously unwelcome to the metaphysical apologists of "all or nothing", while proving heuristically productive in more pragmatic disciplinary contexts (architecture, sociology, etc.). It even comprises Ludwig Klages' genialogical arbitrariness: he declared the *genius loci* of Munich non-decomposable and not very visible, but then traced it back to quasi-objective although variously interpretable conditions (Bavarian geography and climate), and above all to two evocative and imaginary symbols like the rusty exhalations of the cellars and the Marian blue of the sky (not surprisingly taken from various urban elements).

> No *Genius Loci* can be broken down, as it is the synthesis and the essence of largely invisible elements [...] And for this reason, the elements of the soul of the place that we must mention include: the Bavarian plateau with its lakes and its moors, its impetuous mountain streams, its bright midsummer days, the harshness of its air, the autumn mist, the sultriness of the föhn gusts and the thundering storms. We have listed some conditions, but the soul of the place as such does not lie in these things, it

> cannot be expressed: one can only look for its symbols, successfully or not. We have identified two of them: the rusty exhalation and the deeply diaphanous blue of the sky. Symbols [...] are active realities and active realities also always leave a trace in the material reality of things: the Bavarian national colours were white and blue; post boxes, tram rails and uniforms were blue; blue appeared in the "Munich emblem", in the incomparable Frauenkirche, in the vaulted ceiling of stars; the cloak of the "mother of God" is or should have been blue [...] But the rusty exhalations, which imperceptibly vibrated in the night above every brewery and above the chestnut summer candles of every cellar, formed a dense cloud visible in the open sky, above the human confusion of the night, the noise, the mist, the bright sparkle and the glow of every Oktoberfest.
>
> (Klages, 1940, p. 18 f.[16])

Klages's poly-iconism and poly-daemonism[17] – "there are [...] as many spaces and times as the images it is possible to have a lived experience of, in a space-time sense" (Klages 1929–32, pp. 1263–1264) – suggests that there are as many *genii* as the perennially changeable ecstatic-contemplative experiences of the human being.[18] Indeed, the "et cetera" that inevitably ends the undoubtedly evocative qualitative lists made by Klages[19] and every other romantic apologist of the *genius loci* is a suggestive but theoretically useless poetic leitmotiv.

In contrast to this inflationary prospective-imaginary move, which undoubtedly risks reducing the *genius loci* to the invention of a brilliant and exuberant writer, one would have to evaluate, certainly with the precision allowed by a pathic-prereflexive impression, if a certain place has a spirit independently of one's subjective projection – a spirit that is dominant even in the geomantic approach,[20] and which remains unmistakable even if it is unmentionable (like the place where Rome was founded according to Virgil).[21] The obligatory question, for the scholar no less than for those who design places and environments, is therefore whether the *genius loci* is *found* – following an archaic numinosity (Rudolf Otto) or the total impression perceived by the observer (Alexander von Humboldt) – or whether it is *created*, as obviously required by a certain architectural tradition (from Le Corbusier to Botta, for example), perhaps even by inserting provocatively artificial elements in the place, by way of contrast (cf. Kuhlmann, 1998, pp. 8, 14). The Heideggerian idea of responding to Being's appeal and thus preserving its Fourfold (Heidegger, 2001, pp. 141–159) is very widespread among architects. Instead, significantly, the same cannot be said of the philosopher's contextual insistence on the preliminary need for a releasement towards things, whence the rather controversial argument that, just like the landscape, even the *genius loci* should be the result of constructive initiative, at best favored by the potential that would be sedimented in it.[22]

My proposal, consistent with the idea of three ideal types of atmosphericness, presupposes instead the recognition of the manifold nature of our dialogue with a place. When interacting with a place, even the same one, we can in fact a) abandon ourselves in an ecstatic-cultic way to it; or b) intervene on it through material or ideal superstructures, as with the Christian reinterpretation and architectural refunctionalization of previously pagan sacred places; finally c) we can flatten it, making it perfectly suited to host new architectural projects (almost) totally unrelated to the character of the place.[23] Only in the first case do we have a full experience of the *genius loci*, whereas in the second we re-functionalize the primal (and still active) potentiality of the *genius*, and in the third we embrace the constructivist illusion of being able to generate it from scratch in a space that is evidently considered available and semantically indifferent (i.e. devoid of its own *genius*). One should therefore question the thesis that the experiences of places "are under our control" and can be improved, since we "can simulate them; we can pre-visualize them; we can redesign them; we can manage them; we can add on new awareness" (Hiss, 1990, pp. 99–100), as the *genius loci* cannot be "reduced" to human design.

It is certainly legitimate to state that both the *in situ* perception of the urban ambiance and dwelling, understood as a configuration, modalization and articulation, involve the overcoming of traditional dichotomies (subjectivism vs. objectivism, active vs. passive), and that therefore an atmosphere should not be thought of as a static contemplative "object" but also as a constructive circulation between the given (felt) and the configured (enacted, cf. Thibaud, 2015, pp. 28, 50, 57–58, 227). But maintaining that perceiving an atmosphere always implies an (at least virtual) action is certainly not tantamount to saying that an atmosphere is always and only the result of an intentional project. Even those who emphasize the sensorimotor dimension of the encounter with the world cannot disavow the primacy of the action of the environment. In fact, the latter's rhythmic-energetic order (suggestions, lines of force, etc.), understood as a holistic affective tonality including a subject and an object that are not even properly differentiated yet, is what guides the designer's decision to act in a specific place (and not elsewhere), and especially in the case of the *genius loci*. In other words, the designer pragmatically answers the question "what does this ambiance or atmosphere want and allow me to be, do and perceive?". Even if mediated by diffused, unanalyzable, infraconscious and unfocused *petites perceptions*, the *genius loci* in my opinion is an atmosphere or "pervasive quality" (Dewey) that, due to its particular intensity, involuntary nature and authority, is something relatively unavailable: it impregnates a place, demands to be "heard", and suggests design and dwelling performances that delude us to think we are autonomous designers of atmospheres. If there were only this prototypical atmosphere, refractory to any attempt at subjective transformation, "setting up a *genius loci*", with its teleological and semiotizable implications, would even sound like a *contradictio in adjecto*.

But the projectivistic-constructionist conception also has a long history. It radicalizes both the homeric universalization of divine characters, merged into olympic-panhellenic divinities that, unlike local daemons, can be incarnated in landscapes (not necessarily the original ones),[24] and the subsequent monotheistic disavowal of the spirits of places. In fact, interrupting the mythical continuity and transforming the protective daimon into a tempting daemon, enthusiasm into diabolical possession and pilgrimage into a heresy compared to the illocatable transcendence of the divine,[25] Christianity exorcised, exiled or at least degraded the local spirits (which, contradictorily, were still considered existing). Thus, Christianity produced a topophobia for which even the places that under popular pressure were later regarded as dispensers of grace and therefore worthy of pilgrimage – caves, cliffs, trees, springs, mountains, provided they were duly purified – are at most allegories of celestial Jerusalem, whose potential sacrality depends exclusively on some extranatural miracle that left some trace there. Regardless of the potential setbacks (European paracelsianism, Romantic pantheism, qualitative geography), this easily leads to post-Enlightenment subjectivism, according to which the spirit of the place, first understood as numinous (pagan antiquity) and then as inspirer of human (Renaissance and Romanticism) also aesthetic *genius* (cf. Grassi, 1979, p. 187 ff.), must be replaced by the idea of place as "white canvas" on which the emancipated subject could arbitrarily project their own meanings: an idea whose climax was probably the voluntaristic-avant-garde atmosphericism sought by situationist psychogeography.

My hypothesis is therefore that a place has its own *genius* only if (when and where) it radiates an intense and authoritative specific atmosphere. And this *genius*, as we have seen, is essentially found, and perhaps sometimes produced, but only within certain limits. Borrowing and adapting the tripartite spatiality proposed by Hermann Schmitz (space of wideness, directional space, local space), one could also say that the *genius* sometimes presents itself as a semi-climatic atmosphere of entire regions, sometimes instead as the atmosphere of a circumscribed landscape (a mountain, a valley, a bay, a city, etc.) and sometimes even as an atmosphere condensed in a particular place (a tree, a cave, a road intersection, a building, etc.) or in a single object, which are likely to act only as spheres of condensation of wider atmospheres. Only this inflationism makes it possible to explain why the *genius* of a certain place seems to be only circularly ascertainable, in the sense that, on the one hand, unspecified atmospheres seem to materialize in certain places, and on the other, certain places propagate their atmosphere in larger territories. That is why the *genius loci* of Paris can legitimately reside both in its overall *Gestalt* (perhaps even reduced to clichés) and in some features that are only salient for the perceiver's biography (that bistro, that square, certain perfumes, etc.) and are maybe the result of a momentary and unrepeatable sensation.

> A hundred times I must have seen the forest in front of my window, experiencing nothing but the thing, the same thing that even the botanist has in mind: but once, while it was burning in the splendor of the evening sun, its sight managed to tear me away from myself: and then my soul suddenly saw what I had never seen before, perhaps for minutes, perhaps for seconds; however, for a short time or a long time, what I glimpsed was the original image [*Urbild*] of the forest, and *that* image will not come back for me or for anyone else.
>
> (Klages, 1974, pp. 422–423)

Despite this fascinating but heuristically unmanageable "profane illumination" (Benjamin), one should recognize that every expressive variation remains mostly anchored to atmospheric and genial types (first of all natural but also historical) that are so relatively objective as to be repeatable and widely shareable. Alexander Pope's invitation to consult the Genius of the Place everywhere is therefore equivalent to the invitation, consistent with a pathic and atmospherologic aesthetics, to abandon oneself as far as possible to the quasi-thingly nature of the genial-atmospheric qualities immanent in our "surroundings". With this aesthetic-phenomenological re-evaluation of the *genius* (of some and not all places), largely converging with the "geopiety" of much qualitative geography,[26] I therefore intend to distance myself from late Romanticism, often degenerated into obscurantist and chauvinist localism, and from post-Enlightenment isotropy, so contradictory that even the ontological "nudism" opposed by Levinas (with the praise of the homogeneous space symbolized by Gagarin) to Heideggerian topophilia[27] had to come with the messianic appeal to a place, evidently perceived as more "genial" than others (enough to cause war!), like the Holy Land.

As we have seen, the theoretical thematization of the *genius loci* is always a little unsatisfactory under the strictly epistemic profile. However, provided that the intervention aimed at the preservation and/or amplification of the spirit of a place (altars, temples, nymphaea, gardens, monuments, iconification, edification, etc.) is not confused with its arbitrary projective production, it seems legitimate to affirm that the *genius loci* – at least in the sense of the prototypical atmosphere – is not intentionally constructed. Rather, it is unintentionally found in the lived space as something "given" that is "brutally there", as Dewey says of the total pervasive quality (cf. Dewey, 1931, pp. 93–116). In short: despite not having a clear-distinct idea of the spirit of a place, we cannot help having a clear-confused idea of it (therefore an exquisitely aesthetic idea, according to Baumgarten) when deciding where to live: which proves, *pace* Kant, that with the *genius loci* something may be true in theory without applying in practice.

Notes

1 Think of the control policy inaugurated in the Napoleonic era, consisting of depersonalizing houses by replacing their names with mere house numbers (cf. Pennick, 2009, pp. 64–65).
2 As surprisingly happened even in an "urbanized" philosopher like Socrates (*Phaidr.* 238 c-d). The Greeks, however, lacked a technical term that perfectly corresponded to the Latin *genius loci.*
3 The city of Wiesbaden would probably have been related to the element of water (Neddens, 1986, pp. 109–110) even by an urban planner uninterested in the *genius loci.*
4 Here it is easy to go wild: from the magnetic-radioactive radiation of certain areas to Chinese feng shui as a formal adaptation of housing to the landscape, and so on (cf. Brönnle 2009).
5 For the criticism of frontality cf. Pallasmaa (2005).
6 This socio-climatic explanation of this character should rather be supported on a symbolic level, which is both fascinating and arbitrary: for example, a visionary and gnostic like Alfred Schuler, in his speeches on Rome as "the eternal city" (from 1915 to 1923), sees the external urban space as a gigantic interior and therefore as the original religious fact (Schuler, 1997). For a disenchanted and in some ways healthily antithetic (negative) perspective on the *genius* of Rome, see, instead, Warhol (1975, p. 157), for whom Rome would be "an example of what happens when the buildings in a city last too long". For him, Rome is called the "eternal city [...] because everything is so old and everything is still standing. They always say, 'Rome wasn't built in a day.' Well, I say maybe it should have been, because the quicker you build something, the shorter a time it lasts". After all, "buildings should be built to last for a short time".
7 Where we "return to the origins, and rediscover the 'idyllic' world of the forre and the vernacular settlements" (Norberg-Schulz, 1980, p. 164).
8 Thus typology is curiously similar to Claus' (1926) geographic-anthropic and far more deterministic one.
9 However, (*contra* Valena, 2009, p. 200), I doubt that the *genius loci* is all the more effective the more it depends on a human factor.
10 "The ancients knew about the *genius loci*, the nimbus, the aura, and we still talk of the 'atmosphere' of a person, a house, a place. Now, this 'atmosphere' – which so-called sensitive natures capture and refined ones feel, while remaining unknown to more robust minds – is a reality that acts by giving and increasing or by absorbing and weakening, by wrapping up and heating or by digging and cooling down, by soliciting and exciting or by braking and damping down, by expanding or compressing, by encouraging or paralysing, and its action is essentially different from the touching of bodies" (Klages, 1929–32, p. 1103).
11 Regardless of whether the tonality is only vaguely felt or more precisely perceived in the environment: cf. Augoyard (2016, pp. 579–580).
12 At least in some aspects, this is similar to a "simultaneous perception" (Hiss, 1990, pp. 3–26), or to a peripheral and unfocused perception in the anti-retinal and anti-ocularcentric sense described by Pallasmaa (2011, p. 189; 2014).
13 The specific Greek topography (well-defined sites) must have lent itself well to the identification of places and their *genii* (see Norberg-Schulz, 1980, p. 28).

14 Cf. Garnham (1985, p. VIII, p. 9): this purely defective feeling should not be underestimated, however, since it always signals a preferable status (be it real or imaginary).

15 This takes nothing away from the heuristic usefulness of the suggested distinction between prototypical, derivative and spurious atmospheres. My distinction, in some ways, is comparable to that proposed by Kozljanič between a *genius loci* that is objectively immanent to a certain area (and therefore prototypical), a *genius loci* that is culturally objectified, so as to leave traces, and finally a more subjective *genius loci*, nourished even by associative and idiosyncratic projections, and for this reason only responsible for transitory and obviously not intersubjective qualities.

16 This description, although too vague and literary, is much more expressive than that which, for example, attributed to the same Bavarian city five human types, depending on the geological, vegetal and faunal varieties of the surrounding region (France, 1920, p. 271, p. 296 ff.). See Falter (2006, pp. 355–356). It is also ultimately less generic than that according to which, with presumed historical accuracy, the *genius* of Munich (recognized in metaphysical empiricism) comprises romantic, anti-egoic, anti-capitalist, vitalist, movimentalist and even ecologist "characters": on the contrary, these are so far from essential as to be ousted by the current domain of finance (Falter, 2009, p. 104). Also, on closer inspection, they can be deduced from a history of ideas disinterested in the *genius loci*: hence the rather unlikely effort to contrast the rusty exaltation (re-emerged in Nazi brown shirts!) with the strength of celestial blue (re-spiritualization) (see ibid., pp. 101–102).

17 "The true daemon is the daemon of a place, a district, an element; changing along with the manifestations of those, it accepts the sacrifices of its followers according to a temporal rhythm" (Klages, 1929–32, p. 1264).

18 See also Klages (1952, p. 46 f.; 1944, p. 259 f.) who in fact attributes a *genius* to plants and springs, day and night, rain and sun, threshold and city, entrance and exit, etc.

19 More than the metaphorical-aesthetizing oblivion of the "reality" of "original images" (reprimanded to Klages by Kozljanič, 2004, II, pp. 297–299).

20 By geomancy I mean here not so much an esoteric-divination practice but rather a more rational integrated science of space, atmosphere and *genius loci* (Müller, 2009, p. 142).

21 "Hoc nemus, hunc, inquit, frondoso vertice collem, / (Qui Deus, incertum est) habitat Deus" (*Aen.* VIII, 351–352). "'These wooded steeps', he said, 'this sacred grove / What godhead haunts, we know not'" (transl. E. Fairfax Taylor).

22 The *genius loci* would be "determined by what is visualized, complemented, symbolized or gathered" (Norberg-Schulz, 1980, p. 58).

23 Cf. Kozljanič (2004, I, p. 387 ff.; II, p. 117), who, however, codifies this typology a little too rigidly.

24 But evidently considered more able than others to radiate that specific sacred atmosphere, that is, a given ontological characteristic. Hence the erection of temples dedicated to them in various places. Cf. Philippson (1939) and, for an analysis in this sense of the various landscapes of Apollo, Hera, and Zeus, cf. Kozljanič (2004 I, p. 160 ff.)

25 For the first reformed anti-daemonological rationalism, in fact, when one place appears more sacred than another, it is legitimate to suppose that there is some diabolic action at play. Cf. Kozljanič (2004, I, pp. 285–295, 402–407).

26 Cf. Tuan (1976), Relph (2009) but also Ingold (2015, p. 87). The latter, in the context of a reversibility of the "flesh" *à la* Merleau-Ponty, sees the meshwork as "exhalation" and the atmosphere as "inhalation", that is, as a kind of reactive passivity.

27 To the enigmatic fascination of places and natural landscapes, which would be responsible for an idolatrous and somewhat pagan infantilism fraught with cruel consequences, Levinas (1990, p. 231 ff.) contrasts the human delocalization made possible by space travel, but already anticipated by the process of abstraction and disenchantment of space (thus made isotropic) typical of Judaism.

13 Stone, soil and sand

Earthy atmospheres

I have often referred to atmospheres as sets of expressive qualities (see especially Chapter 7). Now let's try to "test" this approach on earth, that is, on one of the elements that have been downgraded from vital media to threatening obstacles because of the irresistible rise of technoscience,[1] and with respect to which man is either a master/tragic hero or little more than a degraded "comic foil". However, today's serious ecological issue seems to be stimulating a sort of rebirth of natural elements through a "commemorative thinking" (Heidegger), leading to their "return" to human lifewordly awareness. Common sense, largely colonized (and induced) by modern physics to focus on the microscopic and macroscopic components of the material world, has long underestimated natural elements, which for centuries have been desemantized and discredited for the benefit of an *Umwelt* condemned to an anodyne isotropy.

Today, however, these elements are seemingly reaffirming their vital centrality. This is mainly due to their unquestionable sensible concreteness (which, from my point of view, is *ipso facto* aesthetic) and to their consequent resistance to the anthropic, since, to quote Schillers' *Song of the Bell*, "Elements have ever hated / What the hand of man created" (vv. 167–168; cf. Münsterberg, 1916). The elements are becoming once again central, one could say, also thanks to their felt-bodily effect which, although not necessarily re-mythized in a divine or at least daemonic form, undoubtedly compensates for the aesthesiological deficit of the increasingly unnatural and urban present lifeform. And it does so aesthetically but also artistically, as can be seen exemplarily in certain strongly material-oriented works of art that, by underlining the neutrality and unproductiveness (non-sowability) of the soil,[2] seem to perfectly respond to the intrinsically anti-Enlightenment need to take the elements "as a partner of man in his sensible and felt-bodily existence" (Böhme and Böhme, 2004, p. 302).

But has the earth really been forgotten? Perhaps we do still rely on its solidity in order to build, both in a literal and in a metaphorical sense. Perhaps the fearless peak climber still challenges its resistance. And perhaps, more prosaically, we still count on its physiotherapeutic benefits when using mud and clay masks. Yet it is true that "for centuries there have been no intercessors and exegetes of the earth in philosophy" (Welsch, 1991, p. 23). And when, as in the 20th century, there have been some, they have unfortunately passed off the harshest nationalisms and the vaguest palingenesis as interest in the earth. So much so that it has taken much effort to save the Nietzschean and Übermensch-like "faith in the earth" from the annoyingly regressive flavour given to it by the subsequent historical context, in order to rediscover, more correctly, a phenomenological acceptance of genuine sensible experience. A pathic aesthetics therefore can only enhance the expressive-atmospheric qualities of the earth – as it could and should also do for the other elements – and identify its specific felt-bodily resonance.

But it is easy to say "earth" (or "land!", like the sailor on the lookout), or that one ought to "have one's feet firmly on the ground". In fact, it is absurd that humans "believe they are already glimpsing nature at the sight of a potato field, and are greatly satisfied if some starlings and sparrows tweet on the few trees on the main road" (Klages, 1974, p. 616). Similarly, I am certainly not reductionistically thinking of the mineral composition of the earth or of its economic purposes, nor do I intend to insist too much on its being humus (in a broader and not just chemical-organic sense) or on its mythical character (Gaia). Rather, here I will try to discuss the expressive-atmospheric qualities of the earth (such as stone, soil and sand), escaping as far as possible the Procrustean bed of physicalistic experimentalism (observability of data, reproducibility and intersubjectivity of knowledge, prohibition of metaphysical hypotheses, etc.) and remaining loyal to phenomena, that is to say to the present and quasi-thingly reality in its sensory-moral value (to use the Goethian notion).

The first thing to clarify is the type of felt-bodily communication that is established between the perceiver and the earth. As we already know, we could speak of an alternation of encorporation and excorporation, since, like any perceived form (emblematically the gaze of others), the earth is both a centre of encorporation and a demanding attractor of attention. At this point it should be obvious that the atmospherological approach to the earth would be completely identical to the effort of determining "a beauty intrinsic to material substances; their many hidden attractions, all that affective space concentrated in the interior of things" (Bachelard, 1948, p. 8), were it not for the radical de-psychologization of the affective that the pathic-neophenomenological approach pursues. In fact, the lived experiences that interest me are irreducible not only to physiological and neurobiological processes but also, unlike what happens in Bachelard's masterful material psychoanalysis, to their introjective localization and their subsequent projective externalization. In this sense, somewhat in line with what Sartre says of the material meanings really belonging to things (and, I would add, even more to quasi-things), our atmospheric feeling of the earth – which is ours not because we own it, but because it *affects* us – is an external situation that is largely irreducible to discrete elements. It is indeed a quasi-thing that tonalizes our surroundings thanks precisely to the pathically meaningful impression[3] it makes on our felt body. Likewise, "the yellow of the lemon [...] is not a subjective mode of apprehending the lemon; it *is the lemon*" (Sartre, 1969, p. 603).

Of the three aspects that are often inextricably merged in the polyvocal notion of "earth" – soil,[4] ground,[5] planet Earth[6] – I will initially focus on the second, meaning every surface that remains immobile compared to the rest: an element that is both reliable and yet goes mostly unnoticed. Its potential and sudden seismic instability is in fact mostly traumatic, first of all in the felt-bodily sense (lack of balance, motor uncertainty) and then also in a physical-bodily sense (fall). Thus the atmosphere of reliability normally irradiated by the ground – which for its stable resistance makes any mechanical activity possible, providing it with support[7] – is replaced by a veritable atmosphere of betrayal,[8] fuelled by the tragic awareness that the previous trust was only based on the naive denial of a permanent threat. As the necessary still background of every mobile object, the ground then generates an atmosphere of gravity that induces a twofold state of mind: that is, the desire to free oneself from it to do (*à la* Nietzsche) something noble (the "ascent" or "climb" in both a mountaineering and a professional sense), but also the desire not to lose contact with it, if not to descend (but not sink!) into it, in order to reconcile with one's roots (Mahayni, 2003, p. 184), or even with the tranquillity that represents "the ark of the world" and the "primordial homeland" (Husserl, 2002, p. 130) – indeed, for Husserl the Earth neither properly moves nor stays still. By transforming every human being into an Antaeus threatened by the slightest detachment, the primary atmospheric nature of the earth's soil serves as a perennial warning to stay, as it were, "connected to the earth", even in a juridical sense,[9] albeit through very different motor suggestions and synaesthetic impressions – depending, for example, on whether you are lying on the ground or getting up, immersing yourself in it or being swallowed by it, and so on. Instead of further describing the semantic values of the "ground" or "foundation", which have always haunted philosophers, I will now try to clarify, following the example of the ecstasies of *every* material,[10] the specific affordances of some different types of earth,[11] i.e. what they suggest and how they do so.

a) *Stoniness.* The first impression is that stone resists us and is hostile to us, being both helpful and an obstacle. In its icy solidity, it is almost antithetical to the maternal dampness of the soil, being the alchemical *caput mortuum* or the allegorical skeleton (acting as a *memento mori*) of the living mother earth. However, stone is nevertheless aestheticized in various atmospheric qualities. First of all, it is so hard and impassible as to express eternalization, whence its usefulness when, as in funeral monuments and classical sculptures (cf. Mahayni, 2003, p. 203), we try to escape from the dissolving force of time, at least relatively (erosion happens, but certainly over geological times that are inaccessible to our perceptual experience, even considering several generations). Being apparently non-deformable, in spite of any external moulding force, stone also suggests a monadism so "wordless"[12] as to seem completely autotelic and consequently radiates an atmosphere of refractoriness, estrangement, and sometimes even of conflictuality that is still a possibility of felt-bodily communication.

At the top of the scale of (natural) hardness,[13] somehow saying "no" to the percipient at first (Mahayni, 2003, p. 191), stone generates an atmosphere of conflict around it, whose felt-bodily pre-reflective resonance is given by a certain anger and by the desire to overcome its dull resistance with one's physical and intellectual strength.[14]

> Reason can very well say that rock is still, sight can reassure us that stone is always in the same place, experience can well teach us that the monstrous boulder is a tranquil presence: the provocative imagination [in my terms the affective and felt-bodily involvement generated by atmospheric irradiation; author's note] has already started the fight.
>
> (Bachelard, 1948, p. 191)

Stone thus spreads a kind of terror in the landscape, inevitably emanating a heroic moralizing atmosphere, which is answered by fighting – a contrast that gives strength to human pathological fragility[15] – everything that "challenges any penetration, any scratch, any wear" (ibid., p. 203), but also, in the aftermath of the 18th-century "discovery" of the Alps, by aspiring to conquer the great peaks,[16] which have now become, if not idyllic, at least picturesque.

The atmosphere of stoniness, in short, is made of a balanced tension and constantly oscillates between agitation and peace. The *Stimmung* aroused by the crushing mass of the Alps is in fact not only antagonistic and irritating, in its apparent figural chaoticity, but also pleasant when this mass appears unitary in its colossality,[17] when, suggesting to the percipient an "induced verticality" and the consequent "panoramic onirism" (Bachelard 1948, pp. 358, 380), it fills the climber with pride, transforming them into a contemplator. The atmosphere of familiarity and safety typical of the earth in general is predominantly overturned, in the case of the earth-stone, into a threatening atmosphere of (non-quantifiable) heaviness – think of a boulder looming over you – whose felt-bodily resonance can only be contraction and tightness. This is also because stony heaviness, evoked in painting by colours like rust or, in fact, "earth" (cf. Mahayni, 2003, p. 206), rejects any material penetration – the interior of the stone is no different from its outside, therefore crushing it would only reiterate on a smaller scale its initial impenetrability. And it also rejects any "calculating importunity" (to paraphrase Heidegger, 1971, p. 46), as the weight measured on a scale has nothing to do with atmospheric heaviness, just as air vibrations have nothing to do with colour. And this can leave the particularly sensitive observer petrified as if they saw Medusa.

Goethe once said that "stones are mute masters, they make the observer keep silent, and the best that one can learn from them is not communicable" (Goethe, 1981, VIII, p. 476). This, however, alludes to the potential esoteric penetration of their secrets. With a more pathic approach, one could say that "the contemplation of the stony petrifies" (Mahayni, 2003, p. 199), generating an evil hypnotism. At best, this atmosphere of impermeability and extraneousness arouses, as said, a handicrafty and extroverted dynamism in the percipient,[18] or a hardening of one's character to resist otherwise intolerable psychic fluctuations.[19] At worst it fully paralyzes the perceiver,[20] sometimes acting as a *memento* of the terror experienced by nature itself, petrified because of the aggressive human indiscretion.

b) *Soil.* In these pathic-atmospherological considerations, the tautological relationship between noun and adjective is not at all a logical *monstrum*, since this relationship expresses precisely the fact that "the substance realizes *its* quality, it [makes us] experience the ownership of its particular richness" (Bachelard, 1948, p. 303). Therefore, if stone is stony, then earth is earthy: being more ductile and porous, more heteroplasmable and less over-temporal, and less of a "thing" than stone, it irradiates a less antagonistic atmosphere. In its general shapelessness, brilliantly exploited both by "land art" and by the less colossal "art in nature", the earthiness of the soil suggests both the genesis of things (whence its central role in the myths of every people) and the mnestic potential of a substance that, precisely because of its humidity,[21] easily collects human traces. The felt-bodily communication that derives from it is therefore probably dialogical: precisely because of its texture, open to all possibilities, earthiness does not arrogantly challenge the agent (like stone), but gently solicits their modulating creativity (Mahayni, 2003, p. 194), and causes them to leave a trace. This is why Japanese Gutai art is able to transform into works of art the traces of its "struggle" with the earthiness of clay, destined to rapid consolidation thanks to its mixture with lime and cement (cf. Wagner, 2001, pp. 114–115).Or else, to make a very different example, this is why we can observe footprints on the ground, be them our own (perhaps to guarantee that we did indeed walk there, as emblematically happened on the Moon), or the traces of people who "returned" to the earth by giving their life, turning into the earth itself (cult of martyrdom land).

It is no coincidence that, being a genuine synecdoche by contact, soil has always been a relic of epochal events. For example, it testifies to the field on which a battle has been won or lost, or – according to the traditional religious and political-monumental *topos* – it testifies to a land whose translation in another place also allows for the latter's consecration. Of course, in order to provide the right context for a relic or to sanctify a new building,[22] this soil must come from the *loca sancta* (for example from the Holy Land); to legitimize a monument to the homeland, it must be soil that the exiled has brought from their homeland (cf. Raff, 2008, pp. 106, 120 f.): in any case, it is an apotropaic *pars pro toto* with a relevant artistic variant.[23] Soil ultimately generates two atmospheres: on the one hand, that of primeval fertility, because "fertile soil contains within it, within the womb of its fecundity, an inner measure" (Schmitt, 2003, p. 42); on the other hand, that of a matter that acts as archive and sepulchre, as all material things sooner or later dissolve into it, thus giving life to very stimulating layers for archaeologists and geophysicists.

3) *Sandiness.* Easier to manipulate than earthy earth, sandy earth is always anthropizing in bearing the memory of "firm lines, whereby defined divisions become apparent" (Schmitt, 2003, p. 42). By its very nature, it "betrays" every stable formal solution, which becomes partly perceptible only if generated by what contains the grains – which as such are discrete and, by definition, incalculable and irrelevant to the whole. Thus, unlike stone (which is always individual, even when it is part of a group), sandy earth obtains its salience only as a collective matter (Mahayni, 2003, p. 196). In its "softness" it radiates an atmosphere of tacit and deaf rather than direct and explicit hostility,[24] a disturbing atmosphere of almost fractal caducity and randomness (cf. Soentgen, 1997, p. 127 ff.) – perfectly summarized by the frustration that comes from writing "on the sand". This atmosphere is promoted by a felt-bodily communication with the percipient whose epicritical points, stimulating various felt body isles, actually generate the "pleasure of reacquiring one's physical body" (Mahayni, 2003, p. 196), precisely because the latter may overcome the annoying point-like impression of the grains.[25]

But the atmosphere of sandiness suggests not only the uncanny sensation of the transience of human beings, whose historical memory seems to vanish just as a footprint on the sand is erased at the first breath of wind,[26] but also transcendence, especially when the sand is presented in its collective form (desert, steppe). The felt-bodily communication, resulting from an atmosphere of timelessness and non-iconicity, induces the percipient to humbly bow before something that, providing no means of orientation, does not seem human. In other words, the perceiver is urged to bow in front of something numinous, which points to the afterlife. In the ideal-typical cosmic-desert landscape (see 3.4), in fact, the perennial immutability of the sky somehow "destines" the human being – more than mountains and more than abstract snow fields (Simmel, 1993) – to monotheistic faith (cf. Norberg-Schulz, 1980, p. 45).

As we have seen, despite being expressed with different degrees of intensity, the atmosphere common to the three types of earth examined is that of a solitary extraneousness, whose felt-bodily resonance in the percipient is often silence. And it is probably for this reason that Heidegger, in his essay on the work of art, sees earth as self-closure, though not without risks and inconsistencies.[27] In other words, earth is the material to which the artist is subordinated.[28] As a *plurale tantum*, it represents the specific trait of the work of art, but it is also the material that, resisting any indiscreet penetration, maintains in its appearance its own irreducible density. Adopting the Heideggerian thesis of the earth's "self-secluding self-disclosing" – one of the few abracadabras that are the pleasure and pain of Heideggerians – one could interpret it as the idea that sensitive qualities, emerging in art in their most specific atmospheric expressivity, neither break away from what they emerge from nor exhaust it, thus preserving its potential. After mitigating the Heideggerian (late Romantic, whether one likes it or not) emphasis on art, the point is simply to reiterate that, like any other expressive-atmospheric quality, even earthiness (now in a broader sense) is a sort of quasi-thingly gaze that the world addresses to the percipient, interrogating them. For Heidegger,[29] expressive qualities thus turn from subjectively designed possibilities projected onto reality into possibilities radiated by reality itself. Stripped of any emphasis on art as the origin "of a people's historical existence", understood as "the transporting of a people into its appointed task as entrance into that people's endowment" (Heidegger, 1971, pp. 74–75), this transformation seems to be perfectly in line with the present aesthetics-phenomenology of atmospheres as feelings and quasi-things that invite us to act, but, as we have already seen, also and above all to (better) feel and perceive.

This reference to perceiving seems to suggest, however, a peculiar relationship between the expressive-atmospheric quality of the earth and the perception of the landscape, duly transformed from purely distal (ocular-contemplative) to deambulatory (felt-bodily and synaesthetically participatory). Undoubtedly, in fact, the *qualia* irradiated by the earth also contribute to the landscape's atmosphere,[30] depending on the impressions aroused by its nature (rock or sand, grass or asphalt, etc.), the way in which it can be traveled, and above all its chromatic tones. Precisely thanks to its colours, the landscape can in fact be relaxing – for example thanks to the polychrome harmony of green, blue, white and yellow – or irritating, for example when the prevalence of a uniform gray-brown shade (symbolic of "bad weather") is so depressing as to push one to go "south", that is, towards different colours of the earth.

Finally, as for the objective or subjective character of the earth's atmosphere, it will suffice to reiterate that if the percipient empathizes earthiness in a certain way, it is only because the earthiness itself pushes them to do so. If, for example, the perceiver senses stoniness in themselves, it is ultimately because they have felt such stoniness as a quality inherent in the place and relatively independent from them. Even in land art, which in its kind empiricism shows an implicit yet substantial attunement with so-called deep ecology (cf. Böhme H., 1992) and perhaps only conceptually alludes to it,[31] the atmosphere of the earth, whose loss is feared by the allusion, is not lost at all. In fact, in this last case the earth suggests neither original strength and fecundity nor a timeless and threatening strangeness and coldness, but a delicate vital balance (perceived to be in danger). And the atmosphere of the earth suggests this by virtue of its immanent quasi-thingly *qualia*: certainly neither because this expressive-atmospheric quality is attributed by us from the outside, perhaps with a "little help" of the transcendental self, nor because it should be sought in a purely semiotic reference of the sensible.

Notes

1 It is no coincidence that Cesare Ripa, in his moralizing allegorism, already chose to represent the four elements as forces that are no longer salvific but destructive and catastrophic (Böhme and Böhme, 2004, p. 274), describing the earth exclusively through the devastating power of the earthquake.

2 To give just one example, see the 197 cubic meters of earth that form *Earthroom* (1977, New York), an installation by Walter de Maria.

3 For an atmospherological phenomenology of the "situation", cf. Griffero (2009).
4 As a substance that can be moved and partitioned, and which is fruitful and welcoming (whence the infinite analogies, mythological or not, with the female body).
5 Be it natural or not: a plane "lands" even if it really only touches the asphalt of the track!
6 Which is still the guiding foundation (for now?) of human life, despite the narcissistic wound inflicted to it by the Copernican turn.
7 "In a swinging boat it's hard to chop wood or to hammer a nail" (Schapp, 2004, p. 22). Cf. Mahayni (2003, p. 183).
8 This is the "traitorous soil" mentioned by Goethe (1981, XIII, p. 257). This explains why every vibration of the earth has always been considered a sign, if not a revenge, of transcendent forces (whence the doubt preliminary to every theodicy) or in any case of catastrophic-epochal changes, physical but also political.
9 In fact, as a place of occupation, of edification and establishment of borders, the earth represents, at least in pre-globalized epochs – before the human conquest of the seas and the airspace – the very destiny of law and peoples (Schmitt, 2003, pp. 45–46).
10 To make an example: the stone suggests heaviness-compactness, metal shimmering-compactness, water an elusive flowing movement, the air impalpability, etc.
11 The earth "is not a natural material even if it appears so to artists; rather, it is thoroughly historical" (Adorno, 1997, p. 148). However, the important thing is not to exaggerate this historicist warning, ignoring as far as possible the earth's historical-cultural qualities for the benefit of its transcultural phenomenological qualities.
12 According to Heidegger (1995, pp. 177, 204–205), this makes it impossible for us to transpose ourselves into a stone (except for cases of mythical-artistic animation). Unlike the animal, poor in world in the defective form of having, the stone would simply be wordless (cf. also Derrida, 1991, p. 50).
13 "With the word 'hard' the world expresses its hostility and, in response, the reveries of the will begin" (Bachelard, 1948, p. 64).
14 "Granite is a particular kind of provocation; its harshness is an offense that can only be avenged with weapons, tools and human cunning" (Bachelard, 1948, p. 22). Culturally unused until Goethe saw it as an original matter (*Urstoff*) placed both in the abysses and in the peaks of the earth, and therefore as a symbol (of the beginning and end of things, of the Trinity, of the greatest stability against the mutability of the human soul), granite came to express the atmosphere of strength, harshness and even, despite its ubiquity, an entirely unmotivated nationalist supremacy in the Third Reich (cf. Raff, 2008, pp. 161–189).
15 "The philosophy of *against* must surpass the philosophy of *towards*, since it is *against* that ends up defining man in his need for a happy life" (Bachelard, 1948, p. 63).
16 Mountain climbs replicate and "democratize" the positivization of the medieval *curiositas*, prototypically found in Petrarca's climb of Mont Ventoux as an anticipation of the modern notion of landscape (cf. Ritter, 2010, pp. 131–137; 1974, pp. 141–163).
17 Because it is "the mass of matter, the unbroken quality of which stretches out undivided among the peaks and gives to their intrinsically meaningless individuality a unified body" (Simmel, 1993, p. 180).
18 As, for example, in Novalis, who considers stone to be fruitfully expressive and suggests to creatively obtain an igneous nucleus from seemingly useless stones: hence an evident allegory of artistic work, which is the more valid the more it involves materials that appear inanimate to the "illuminist" observer (Böhme H., 1989).

19 "How one should turn to stone. Slowly, slowly become hard like a precious stone- and finally lie there still and silent, to the joy of all eternity" (Nietzsche, 2011, p. 263 [§541]).

20 As well illustrated by the Romantic motif of the "heart of stone", possibly also as an (anti-capitalist) allegory of the exchange value and of the immutability of social destiny: it is perhaps no coincidence that the rebel Don Giovanni is punished precisely by a man of stone.

21 Here I am recalling the ancient tradition, first of alchemy and then of the *philosophia naturalis*, of viscous dampness (*terra pinguis*) as the guarantor of the indissolubility of the sublunary compound substances. Cf. Griffero (2006, pp. 373–380).

22 As happens, respectively, in S. Croce in Gerusalemme (chapel of S. Elena) in Rome or in the Holy House of Loreto, whose walls were allegedly made with Palestinian mortars and limestone.

23 Whether it is earth applied directly to the canvas to tautegorically express itself (Rauschenberg, Beuys), or to preserve the memory of the fallen (Anselm Kiefer, on whose work *Waterloo, Waterloo, et la terre tremble encore*, 1982, private coll., cf. Wagner, 2001, p. 120 ff.) although through the earth instead of human figures (as in Turner).

24 As Bachelard posits, while acknowledging that sand does not generate the nausea of subject/object indistinction with which viscosity paralyzes the lazy (Bachelard, 1948, pp. 27, 122).

25 This impression was even suggested by the "Pompeian" sandy catastrophe staged by Magdalena Jetelova's installation (*Domestication of pyramids*, 1992, Museum of Applied Arts Vienna), where the sand was artificially stopped to obtain a configuration that is only apparently mobile.

26 Think of the imaginary map of the March of Brandenburg by Anselm Kiefer (*Wege, Maerkischer Sand*, 1980, Stedelijk Museum, Amsterdam), where the sand seems to gradually erase the names of Prussian glory.

27 But consider the difficulties faced by those who, like Welsch (1991, p. 78 ff.), want to avoid the relapse into mythical-dualistic thought by explaining the work of art as a conflict between world (openness) and earth (undisclosable).

28 "The fibrous and smooth aspects of things, their brightness and opacity, their rigidity and flexibility, the more or less uniform diffusion of colour on their surfaces, are always modes of the earth" (Welsch, 1991, p. 30).

29 It does not matter if it happened with the turn after *Being and Time*, or if it can already be found in the final parts of this same work (Welsch, 1991, pp. 48–49).

30 For whose irradiations (also physico-chemical) in the geosphere cf. Hellpach (1977^8, p. 158 ff.; 1939, p. 50 ff.).

31 See the installation *Der vertikale Erdkilometer* (Walter De Maria, Documenta VI, Kassel 1977): a metal shaft stuck one kilometre deep into the ground. As a reverse monument, it alludes to the underground space as a necessary foundation, but at the same time it allows us to experience our equally indispensable point of view (cf. also Seel, 2005, p. 96 f.).

14 Like leaves in the wind

Does democracy have its own atmosphere?

It is well known that what I have called "atmospheric games", resulting from the different shapes of an "in-between" preceding/connecting subject and object, always have strong collective and therefore also political effects. Nevertheless, it seems extremely tricky to make an atmospherological exercise on the political scope of atmospheric feelings. The apparent elective affinity between totalitarian regimes and management of feelings, in fact, has always made it rather difficult to talk about atmospheres and politics – especially democracy. Therefore, the lack of in-depth studies on political atmospheres is not surprising. However, any proposal enabling a phenomenology of atmospheres to be combined with democratic politics is worth considering. Indeed, to mention Le Bon (2002, p. xv), even in democracy "men never shape their conduct upon the teaching of pure reason", rather uncritically giving into the impressions produced by images, words and formulas. Also, often politics no longer communicates contents or a sense of belonging – increasingly mimetic and unthoughtful – but rather lifestyles, acting almost exactly as marketing and infotainment. This already applied when people voted by ideology, and certainly applies *a fortiori* for today's undecided opinion voter, who "is not the thoughtful 'independent' he is often pictured" (Packard, 2007, p. 173). When, like in aesthetic capitalism (cf. Böhme, 2017c), show-values overshadow use- and exchange-values and convert need into desires that in principle cannot be satisfied, when free time is fully colonized by shopping and *Erlebnisse* as such,[1] when, lastly, old and new media have the power to produce and control collective pathic effects,[2] whether aimed at obtaining a demagogic agreement or generating an *indignados* movement, there is without doubt an ever increasing tendency to just sell atmospheres (here understood in the worst sense of the word) rather than well-formulated ideas.

As we know, in the context of this seemingly unstoppable politics-media coalition, the contemporary global governance crisis often ends up being replaced by alarmist and sensationalist "breaking news" (cf. Milev, 2012), whose inherent emergency nature produces an anaesthetized and therefore politically manipulable situation which one rightly could call "air design" and which, despite being the normal consequence of the artificial bond between members of any organization (cf. Borch, 2011, pp. 33, 39–40), is today functional both to increasing consumption and to what could be called "atmoterrorism" (Sloterdijk, 2012). However, it still appears hard to engage in a serious atmospherological reflection about politics: and the situation is not exactly improved by the ambiguity of a language use that, depending on context and expectations, euphemistically defines a good political atmosphere either a success or a failure to reach agreement. The academic fear both of over-generalizing the notion of atmosphere as such, and of dealing with an aestheticization of politics that appears to be a matter of political science, might also help explain this gap.

However, I believe that, together with so-called economic psychology, a pathic and atmospherological aesthetics, based on the externalisation or de-psychologization of affects and emotional life, would not only make it possible to better understand otherwise unintelligible collective phenomena such as fluctuations in exchange rates, public opinion and electoral consensus – difficult to trace back to conventional contractual transactions and based on a rationality that is only interested in calculating benefits and losses – but would also be able to provide us with a critical atmospheric competence. By working more closely with phenomenology and aesthetics, for example, political thinking could and should firstly investigate the moods and atmospheres of the public opinion, namely the climate or collective affective intentionality prevailing in a certain *milieu*, which, precisely by acting as a latent and non-thetic background, through a complex felt-bodily communication based on motor suggestions and synaesthetic characters influences the specific felt-bodily resonance that tacitly guides behavioural patterns and values (even phantasmatic) of a given community. A depressive megalomania or a disenchanted serenity, mere legal formality or a deeply involving civicness, simple geopolitical belonging or a truly shared destiny, reasonable hopes or self-inflicted complaints: all these and other conceivable conditions are also, indeed, real atmospheres – or at least atmospherically conditioned moods.[3] While they force politicians to decide, more or less consciously, whether to arouse or silence them, they can be studied and perhaps amended (within certain limits) only atmospherically. Although not arbitrarily producible, contrary to what relativistic constructionism claims, a collective atmosphere can only partly be produced, in fact, through a mix of sub-atmospheric generators,[4] including expressive qualities or affordances immanent in the given situation[5] and already settled felt-bodily habits that deserve closer examination.

First, it should be stressed that, like any other atmosphere, the political one cannot be considered a simple transmission of signals: rather, it is a feeling poured out into a certain (pre-dimensional) space and for this reason, in its quasi-thingly nature, it is responsible for an ever-specific felt-bodily communication. The first goal of a government policy, but also of every executive group, is obviously to suggest a syntonic emotional tone, in the worst cases even by using the so-called small talk and baits[6] aimed at supporting previous assumptions: think for example of the phantasmatic family which every fanaticism is based on (cf. Bronner, 2009), or the hasty and allegedly exemplary decisions greatly facilitated by forgetting source and even message, pluralistic ignorance and background noise (too many or too few interlocutors). Exactly as in the case of packaging, which relies on a completely irrational transfer of qualities from package to content, political effectiveness, for better or for worse, counts on atmospheric feelings which could be contagious enough to prevent any probable discrepancy between the starting feeling or first impression and the following demystifying one. When, for example, a sudden firing unmasks the optimism underlying the myth of the invisible hand or a major *gaffe* exposes the fictionality of the so-called streaming-democracy, what generally happens is that the retroactive focus on previously unnoticed inconsistencies can give life to a deep and violent "unpersuasion", which is, anyway, an atmosphere itself, albeit one of cynical disenchantment.[7]

Of course, as show-values are philo- and ontogenetically embedded in human beings, it would be better to leave aside any Romantic illusion of a real authentic and transparent communication and, instead, be satisfied with making good use of appearances. One needs to admit, in fact, that show-values were also present in ideological affiliation and belonging vote, and that the theatrical and spectacularized model – which can be traced back to the Baroque noble tradition of *theatrum mundi* (Früchtl and Zimmermann, 2001), if not to the Augustan spectacularization (cf. Zanker, 1987) – also explains much of past politics. Today it simply uses more widespread and iconocentric-technological tools: such an everyday aestheticization seems indeed necessary especially because of the increasing aesthesiological untranslatability of many global and future problems (for instance, how could the future energy crisis or a symptomless epidemic be sensibly perceivable?) as well as of the increasingly aggressive mentality of a majority system especially based on the epidermal rejection of opponents. If properly understood, however, even the present theatricalisation-atmospherisation of politics could maybe, as in the past (cf. Münkler, 2001, pp. 152–153), soften its power and unexpectedly increase the weight of public opinion, thus showing, at most, that some political events are now nothing more than a painful and impotent attempt of politics as such to survive to much more influential and insidious power centres (international finance, media, multinational corporations, international bodies, etc.).

It should certainly be recognised that the guiding principles of liberal societies (rationalization, tolerance, egalitarianism, role of public opinion), precisely aimed at eliminating the deceitful atmospheric propaganda of absolutist politics, seem likely to weaken the power of collective emotions and atmospheres, which are the more powerful the less intentional (at the source) and conscious (in the receiver) they are. But this, even in our consolidated democracies, is exactly the reason for the regressive need for an auratically-atmospherically legitimated power – in other terms, of a politics based on leaderism, marketing strategies and emotional designs. Nevertheless, one should also try to rehabilitate democracy from an atmospherological point of view, not only by valorizing its rules and procedures but also by re-discovering its emotional and sensible background. To this end, I would like to point out, for example, the (also atmospheric) positive understatement of anyone who, without harbouring the illusion of fully avoiding any form of rhetoric, openly admits their uncertainties in a democratic society and gives precedence to a policy plan over personal and party interests. It is certainly true that procedures and compromises, typical of democratic management as a *clasa discutidora* (Donoso Cortés), generate atmospheres that are always less intense than the totalitarian or populistic ones. But it is necessary to learn to valorize again the specific democratic *ethos*, along with the values-affects (common good, equality, recognition, individual autonomy, etc.) that might be summed up in the general concept of "openness", understood as the willingness to accept what is new, which is not automatically considered a threat only because it opens up multiple and always questionable options (cf. Ferrara, 2014).

Democracy is deeply steeped in skepticism towards emphatic authenticity, prestige and charisma, and so suspicious of the idea of being represented by superior people as to often convert the legitimating "only they" into a devastating anti-parliamentarian "why they?".[8] I know only too well how difficult it is for such an emotional infrastructure to coexist with the prestige needed by any government and authority. However, without necessarily involving fanaticism and irreversible historical processes needing saviours and/or scapegoats, prestige and credibility should also be found in democracy. According to the neo-phenomenological *Leib*-philosophy, whenever something (a person or a thing) gives life to a unilateral encorporation whose outcome is the partner's almost ecstatical excorporation – even if through memorable places and times, like in cultural conservatorism, or in paranoid ubiquitous prevention strategies (contractual rules, insurance policies, medical instructions, political correctness, privacy protection) like in our deeply alarmistic welfare society – this gives rise to a competent and (fortunately only transitional) charismatic authority, also in democracy. The solution, which of is course easier in theory than in practice, consists in resisting the emotional monopoly of charismatic atmospheres (not only when they have clearly vanished),[9] including the fascinating but undemocratic Far East windy model of power (which allegedly bends the leaves without breaking them…),[10] while, at the same time, being careful not to give carte blanche to atmospheres that otherwise risk to be so ironic as to condemn people to inaction.

Roughly speaking, one should learn over and over again to appreciate even the most superficial forms of freedom,[11] remembering the everlasting appeal of democracy to those who live in undemocratic societies. The point is to appreciate the atmospheric benefit of a discourse that, being in principle non-essentialist and non-Cartesian, is bound to precedents and to the burden of proof, temporally and procedurally limited, and only ready to accept what is rationally convincing, without choosing peripheral paths to elaborate messages in order to save cognitive energy. Provided that democratic ideas are not easily exportable, especially because of their inherent and even perceptual counterintuitiveness in rejecting differences and destinies, they can become genuinely efficient only when they radiate an atmosphere of openness. Mindful of the European hermeneutical wars carried out in the name of one sole meaning – first of the sacred texts and later of history (cf. Marquard, 1989, p. 87 ff.) – this openness teaches us to live with uncertainty and pursuing always perfectible plausibility (not absolute truth). This atmosphere is not special: it is generally already perceivable, for example, in every academic meeting. Indeed, such meetings ensure a regular succession of talks (and opinions) – which as such deny any illusory notion of some truthful perspective from nowhere – and this necessary deferral paves the way to a kind of "inter-patience" (Sloterdijk in Latour and Gagliardi, 2006, pp. 108–110) that perhaps symbolizes the democratic *forma mentis* and atmospherization better than anything else. It is now up to political science to analyse the different kinds of democratic (collective) atmosphere, if possible without using a manifestly inadequate model such as that of the communication between an active sender and a passive receiver, but rather developing an appropriate (even if minimal) atmospheric competence (or intelligence). Only thanks to this competence can one hope to become capable of a) fully feeling atmospheres, b) understanding them – i.e. what an emotional authority, even the democratic one, necessarily consists of – and possibly c) distancing oneself from them, that is, feeling them without being grossly manipulated by them.

What has often been referred to as atmospheric "instinct" or "flair" (Tellenbach, 1968, p. 49) – that is, the ability to critically examine the atmospheres one feels – here should be understood as a skill that, like any other, can be improved through exercise. This competence would enable us to benefit from a "provisional atmospheric morality" (to paraphrase Descartes). However, this is not the place to discuss atmospheric authority, which, to put it very briefly, is absolute when it cannot be resisted and relative when it can be opposed by appealing to a higher level of personal emancipation. I will only say a few words about the alleged or suspected ethical consequences of this atmospherological approach and the risk, often overestimated by opponents of New Phenomenology, of irresponsibly indulging in collective emotions. The first thing to note here is that whatever one may ethically say about atmospheric authority takes a relatively different shape depending on whether, to recall my tripartition, atmospheres are understood a) as objective daemonic powers – external to us, unintended, and with respect to which the subjective component is reduced to the more or less critical reaction of the perceiver (prototypical atmospheres); b) as an external and objective quality, which nonetheless, as the effect of a relationship (however implicit), lies between subject and object (derivative-relational atmospheres); or c), finally, as idiosyncratic and therefore subjective and projective moods (spurious atmospheres).

We have spoken about manipulative atmospheres. However, it is best to avoid any rigid dualism between a manipulative and unethical agent and a culpably manipulated and hetero-managed receiver (Heibach, 2012, p. 263). This dualism fails to recognize that, given the holistic nature of every situation, it is enormously difficult to distinguish between subject and object and thus to definitively say if and when an atmosphere is not manipulated (and therefore democratic). Indeed, one should not ignore the fact that the manipulated person is always co-responsible for their seemingly irresistible involvement, and that bad (manipulative) atmospheres are not only the most artificial and totalitarian ones. There are three points to be made here: a) no one is ever really involved by feelings that one clearly sees as manipulative; b) the manipulative or persuasive appearance is implicit in every practice that generates an atmosphere, even in the most correct one; and c) condemning something as an illusory appearance already implies a different and phenomenologically incommensurable level of "reality", as such inaccessible to the pathic subject, for whom an atmosphere is really such as soon as it is felt by them, regardless of whether it is positive or negative, spontaneous or induced.

And yet I believe that only by acquiring a better atmospheric "competence", by no means reducible to the *affectus non nisi parendo vincitur*, can one really learn to appreciate democratic atmospheres. And one could do so while making room for critical reflection, if this is not allowed by the authority itself (in its best practices) – even more so in today's globalized world, where one should be disenchanted about dangerously anonymous authorities (from the "stock market" to GDP to credit SPREAD, etc.). However, I think it is very important not to build one's hopes up too much about a) the perfect and rational transparency of atmospheric feelings, of which in fact we can always be aware only to a certain extent, as well as about b) the availability of an Archimedean point less fallible than that provided by personal critical sense, about c) the generalisation of the intra-aesthetic paradigm of "suspension of disbelief", which would mean not living the atmospheric experience fully, and about d) the possibility of resorting to a parameter measuring the intensity of an atmosphere, since it is absolutely uncertain whether this intensity is given by the inner peak or the manifest expression of the affect, by its duration or practical implications, and so on. Therefore, one has to be content with a true and proper "provisional atmospherology".

I wish to highlight three aspects in particular. 1) In order to mitigate the objection that, according to my approach, a person would be nothing but a "blind passenger of atmospheres", one should learn how to distinguish, as far as possible, between toxic (which does not mean "untrue") and benign (which does not mean "true") atmospheres, while being aware of walking on thin ice, where aisthesis and ethics mingle. It must be pointed out that toxic atmospheres, which cannot however be reduced to non-atmospheres, are not only those arousing stress and distress but also dissuasive-sedative ones. Through them, many *dispositifs* (in Foucault's sense) aim at defusing any social contradiction with the help both of artificial-conformist attunements (Schouten, 2011, p. 103) and of the inhibiting effects resulting from today's obsessive alarmist demand to regulate every fragment of everyday experience.[12] A good atmospheric competence also consists 2) in accepting the fact that, due to the lack in our post-traditional societies of a paradigmatic place of atmospheric awareness, that is of a situation that may act as a paradigm of every other atmospheric experience, one should rather learn to have as many and different atmospheres (spatial, medial, functional, etc.) as possible and thus to allow the resulting experiences to interact with each other. This can give rise to an affective well-being that, exactly as happens in democracy, depends on a division of powers (affective in this case) that relativizes their impact.

Lastly, a good atmospheric competence should 3) favour and foster those atmospheres where, as happens with a *trompe l'oeil*, an early pathic-immersive step might and should be followed by an emergence phase. An atmosphere could thus be poorly manipulative when stimulating a sequence of this kind. In this respect, not to mention here the case of protests offering real counter-atmospheres, aesthetic experience is a good example of an atmosphere requiring both immersion and emersion – which might sound at least in part like the "the dreamer calling out to himself in the midst of the illusory dream world, but without disturbing it, 'It is a dream, I will dream on'" (Nietzsche, 1999, p. 25). Indeed, this atmosphere, powerfully and influentially contagious without ever being oppressive and coercive, is especially found in contemporary art (Schouten, 2011, p. 106): in fact, unlike the populist atmosphere, which as such is hypnotic-somnambulistic, the critical-artistic one generates through its provocative and irritating impact cognitive and affective discontinuities that always allow the observer to take critical distance while empowering those who experiences them deeply.

As already mentioned, for me the authority of atmospheres exists in the prototypical sense only when it overcomes all the critical scruples that the perceiver may mobilize, namely when it prevails over their resistance and the perceiver cannot access a further critical level. Surely, passivity is neither really a problem nor a taboo for a (neo)phenomenology whose core is precisely emotional rapture. Indeed, this approach does not give way either to an ascetic-rationalistic control of external feelings or to arbitrary projective transformations of them: rather, it allows for a distance from them that is always only relative. Therefore, though the atmospherological insistence on emotions may come with some risks and lead to suspicion in a democratic situation, I hold it to be much less dangerous than the current general illusion that the emotional sphere can be universally controlled and manipulated (today even in chemical-genetic terms). As a matter of fact, the typical outcome of a rationalist-reductionist approach is astonishment and indignation about the fact that a person, despite their alleged proud autonomy, has not been capable of simply saying "no!" to a deeply moving collective feeling. In short: atmospheres should not be left to totalitarian regimes and demagogic strategies! My atmospherological approach especially supports the idea that adults, including committed democrats, do not remove at all costs the passivizing-negative sphere and the felt-bodily influence of spatialised feelings.

Rather, and more modestly, they do not neurotically try to escape from atmospheres but come to terms with them in various ways, not least by trying to make a good and democratic use of them.

Notes

1 "New York restaurants now have a new thing – they don't sell their food, they sell their atmosphere". When people "go out to dinner [...] instead of 'going out to dinner' they'll just be going out to atmosphere" (Warhol, 1975, p. 159).
2 More quickly and inexpensively than architecture and urban planning (Böhme, 2006, p. 171).
3 By borrowing Schmitz's concept of "situation", political scientists-atmospherologists might ask whether the situation that is the object of criticism, or in any case of political evaluation, is a) common or personal, b) ongoing or available over a longer time frame, c) impressive (immediately and totally significant) or segmented (manifest only in portions), d) rooted or merely and provisionally inclusive. Cf. Grossheim, Kluck and Nörenberg (2014, pp. 53–57).
4 For an analogous, partial, acceptance of atmospheric constructivism as the generation of effects due to the proto-atmospheric qualities of the components (which are multiple, including climate, history, economy, architecture, communication resources, prior knowledge, etc.), cf. Grossheim, Kluck and Nörenberg (2014, pp. 19–24, 57–58).
5 "Collective intentionality has to do not only with the (logical) constitutive and functional conditions of an institution called 'Supreme Court', but also with the circumstance that a Supreme Court exists and the *significant fact that it looks like a temple and not like a shack*" (Grossheim, Kluck and Nörenberg, 2014, pp. 50–51; italics mine).
6 That is, formulas, influent images, partial analogies, leading questions, factoids (cf. Pratkanis and Aronson, 2001). The manipulative expedients include, on the one hand, the rationalization of a cognitive dissonance artificially provoked in the recipient who, feeling guilty, uncritically accepts the solution offered by the source; on the other, the activation of phantasmal mechanisms by which a thing is all the more desirable the less it is (believed to be) available.
7 "The crowd [...] considers the fallen hero as an equal, and takes its revenge for having bowed to a superiority whose existence it no longer admits" (Le Bon, 2002, p. 88).
8 "The atmosphere of Obama is tangible: his race, his cool, his vocal cadence, the rolling, relaxed swagger of his gait, the aura of hope which swept him to the presidency, and the atmosphere of resignation which has descended on his presidency" (Grant, 2013, p. 18).
9 Take Leni Riefenstahl's propaganda films (*Triumph of the Will*, 1935, and *Olympia*, 1938) or, *a fortiori*, Mussolini's facial expressions as he harangues the crowd from the balcony of Piazza Venezia: who still perceives an effective atmospheric fascination in these images?
10 See Jullien's example in Latour and Gagliardi (2006, pp. 166–177), in the light of a model according to which, inexplicably, the wind always blows in one direction (from power to the people).
11 "Free countries are great, because you can actually sit in somebody else's space for a while and pretend you're a part of it. You can sit in the Plaza Hotel and you don't even have to live there. You can just sit and watch the people go by" (Warhol, 1975, p. 146).
12 For this criticism of our alarmist society see Sloterdijk (2001–2016, especially the third volume *Foams*).

15 Intentionality and unintentionality in the pedagogical atmosphere

As we all know, neither the square feet of their bedroom nor the financial prospects offered by their family can really explain a teenager's affective and intellectual habitus. And we also know that changing the desk arrangement in a classroom can generate a different emotional situation: for example, the individualistic and hierarchical one typical of traditional lectures where they are placed in parallel, at right angles as well as at a proper distance from the teacher's desk; or else, the situation of peer and seminar-like lectures, when they are placed in a circle and the teacher, getting off their desk (metaphorically and literally), sits between the students, without taking a privileged position. Trivial examples such as these already show that the question of atmospheres is not entirely foreign to pedagogy. And yet, making it the object of a true philosophical reflection, understanding atmospheres as feelings effused in the pre-dimensional space, undoubtedly means undermining today's predominant reductionist tendency even in pedagogy, aimed at "reducing" the educational question to cognitive, constructive or even hastily neuroscientific aspects. The recent "career" of the concept of atmosphere therefore also involves the pedagogical field. But one thing is worth pointing out: when one says, for example, that there is a tense and competitive or relaxed and friendly atmosphere in a classroom, one is not expressing an irreducibly subjective situation in an imprecise and exclusively metaphorical way. One is giving an accurate account of an affective situation experienced by the students in their felt body (not so much in their organs as in their felt-bodily "isles"). They experience it not because they own it or have generated it internally, but rather because they feel grasped by it as if by an external power, which hovers precisely in that particular space and which cannot be kept at a distance, let alone modified completely.

To be sure, the notion of pedagogical atmosphere had already been introduced in the early 1960s, first in an article (1961) and then in a book (1964, published in 2001) by the German philosopher and pedagogue Otto Friedrich Bollnow. Rightly noting that the term "atmosphere" – which he claims to use for the lack of more precise terms – alludes neither to a superficial-illusory dimension nor to a cloyingly harmonious one, Bollnow defines the pedagogical atmosphere as "the totality of the affective conditions and human attitudes existing between educator and child and providing the background of every educational behaviour" (Bollnow, 2001, p. 11). In other words, what he has in mind is a sort of undifferentiated holistic climate, not necessarily conscious nor linguistically precise. However, unfortunately he does not offer a true theoretical analysis of those implications. Nevertheless, albeit only in passing, he touches on three points that should be at the centre of every atmospherological analysis (ibid., pp. 12, 29, 62, 67, 110). He insists on the fact that the atmosphere a) subsists "between" educator and student, who are so mutually implied that they are only relatively distinguishable – in my terms, one could speak here of a relation that is ontologically prior to the related. The second point is that it b) represents a situation in which the external world and the internal world are not yet distinct – in my terms we could say that the mood we call subjective is only the reaction to emotive tones that are primarily external. Finally, he notes that c) the atmosphere, like any authentic *Stimmung* (cf. Bollnow, 1956), neither can nor must be consciously sought and perhaps artificially produced – in my terms it could be said that the prototypical atmosphere can never be the object of intentional construction and planning.

In his writings, Bollnow is always motivated by the existential need to overcome existentialism (excuse the pun): therefore he simply reiterates a conception of life as a progressive extension of the neonatal protection, of the intimacy of the nest-house (cf. Chapter 10, cf. Bollnow, 2011), going through the affective virtues necessary for a good pedagogical atmosphere from the point of view of both the child (safety, cheerfulness, positive openness to the future, devotion, love, trust, gratitude, obedience) and the educator (trust in the potential of others, pedagogical love, operational patience, hope, factual serenity, goodness). Unfortunately, however, he does not seem to be sufficiently aware of the ontologically subversive potential of the very notion of atmosphere. He still appears to be excessively bound to the principle of introjection of sentiments: in his opinion, a serene atmosphere can only be radiated by those who are already serene, and atmosphericness can never precede a concrete intersubjective situation (cf. Bollnow, 2001, pp. 67, 107). Also, he does not feel the need to consider atmospheric perception a rigorously embodied process.

For the "atmospheric turn" made available by New Phenomenology to the various human sciences to be fruitful in its various fields of application, including the pedagogical one, it is necessary to do two things. First, to circumscribe Schmitz's thesis according to which no real atmosphere can be intentionally produced.[1] Secondly, to admit, following the later work of Gernot Böhme, that in certain cases one can at least generate opportunities for, and conditions of possibility of, an atmosphere. More precisely, and recalling my ideal-typical classification between prototypical, derivative and spurious atmospheres, we could say that in the field of education, while the other two types are also present, derivative atmospheres are particularly important, as they are somehow generated "between" master and pupil, "between" the students themselves and, why not, also "between" the students and the material reality with its immanent emotional invitations. The genius and creativity of the educator would therefore mainly lie in providing – with no absolute guarantee of success – situations and materials that can possibly radiate a fruitful pedagogical atmosphere (cf. Wolf, 2015, pp. 8–9). In other words, their ability lies in planning useful atmospheres by the virtue of an intentionality that, however, far from being an interior state of mind, fully coincides with situationally appropriate felt-bodily actions and practices (ibid., p. 14). It is evident that this fine sensitivity and responsibility for situations and the atmospheres that can be generated in them is not too far from what has sometimes been defined as "pedagogical tact".

But what is a pedagogical atmosphere? One should certainly remember that, in the light of the inextricable intertwining between the school world and the more general lifeworld (with obvious emotional repercussions on school life), the educational environment can never be examined on its own. In any case, it can be said that the notion of pedagogical atmosphere underlines how much learning is powerfully co-determined by the emotional qualities of the environment in which it takes place and by the affective tones radiated by the educators. In any form of learning (not just scholastic), one learns not so much "what" but "how" – for example, a modality of "showing" or even "indicating" (in a broad sense) rather than determined notions. For this reason, it is quite reasonable to maintain that the educational process, as and more than in everyday life, is mainly the exercise of a pathic power (cf. Schultheis, 2008, p. 100), which from my point of view is founded on the felt body (and not on the physical-anatomical body) intended as an infallible sounding board of the atmospheric feelings encountered in external space. Whatever the form of the inter-subjective pedagogical exercise, only the affective and felt-bodily involvement that derives from it (cf. Schultheis, 1998) determines the formation of the individual in depth and with long-term effects.

Every time we go from processes that can be perfectly described in the third-person-perspective, which may even detect an affective situation but without necessarily feeling it, to processes such as the formative one, in which we can only be competent in the first-person-perspective, perceiving the situation – that is to say of the states of things, of the programs and problems that make it up – amounts to being emotionally involved. Thus the teacher shows or stages something, and does so with their bodily behaviour and emotional tone, with their gaze and voice (including timbre and tone), with their way of occupying and living space and time, and with their gestures (even involuntary). And the smell, the architectural choices, the objects and the educational forms, etc. which make up the furnishing of the educational place, also contribute powerfully – either converging with or diverging from the impressions concomitantly produced by the human beings involved – to the felt-bodily communication that, as I have often said, is the ultimate object of any neo-phenomenological perception. Hence the deeply interdisciplinary nature of the design of educational spaces.

In its external and internal aspects, in the colours and the materials used, school architecture (in a broad sense) always implies a precise conception of young age and of education, precisely because it creates – for the teacher and above all for the learner – a space unequivocally characterized by a certain atmospheric feeling or mood. Even in its facade, for example, the school building imposes a certain way of experiencing and living spaces, which, in fact, always schematizing a pedagogical intentionality (cf. Schultheis, 2008, p. 108), can be dialogic and egalitarian but also authoritarian and hierarchical. These spaces thus express felt-bodily influential atmospheres that are not always unequivocal, as can be seen from the fact that the same narrow place can be experienced as oppressive and as protective. Similarly, a higher teacher's desk can be experienced as the instrument of an overly intrusive surveillance or of a willingness to communicate through an ideally personalized eye contact. Whether in the form of control and physical-mental discipline or of individual motor and operational freedom, a formative process is therefore always pathic and felt-bodily. To make a different example, another potentially atmospherogenic phenomenon is the whole range of expression inherent in one's eyes (direct or oblique, antagonistic or supportive, etc.)[2] and in one's voice (weak or thundering, monotonous or melodic, etc.): a whole felt-bodily range whose extreme poles are on the one hand epicritical bitterness and sharpness, on the other protopathic softness and diffusivity.[3] But it is not only teacher and student who meet and "feel" felt-bodily: this happens also for the student and the inanimate entities that surround them, by whose expressive qualities – understood, as said, as "ecstasies" of things and emotional affordances – they are more influenced than they think.

In general terms, within felt-bodily communication, the perceived, whatever it is, ceases to be the object of a distal perception and presents itself instead as a true partner of the percipient. In other words, it becomes a subject radiating affordances that turn out to be more or less involving, more or less harbingers of pragmatic outcomes, long before any conscious cognitive or linguistic process takes place. This articulated felt-bodily communication therefore produces an underlying emotional tone (a basic mood or *Grundstimmung*), which partly derives, as I have already mentioned, also from the interaction between scholastic and external atmospheres, primarily family ones (for example, those of coldness but also of superficial harmony are very harmful). This basic mood usually survives for a long time, characterizing for better or worse the whole process of growth of the individual, setting up their lasting felt-bodily disposition. Rather than being thrown into moods or attunements, as claimed by the early Heidegger, one might say, a little less melodramatically, that we grow up in atmospheres as in a "niche", to such an extent that we no longer realize that our moods or attunements are not innate, but rather generated in us by the exposure to a recurring kind of atmospheres.[4]

In the different stages of the formative process, the learner is "touched" in some relatively variable "isles" of their felt-bodily dynamic, which is easily exemplified in breathing as inhalation and exhalation and whose extreme poles are tightness and wideness. The teacher's frowning and severe look, to make a trivial example, is felt-bodily experienced by the student (especially, of course, in primary school) in the form of their own felt-bodily contraction, while a confident and encouraging look is experienced in the form of felt-bodily expansion. Be they negative or positive, these suggestions generate a situation that may be defined an atmosphere, and in which the skilled educator – able to understand the felt-bodily dynamic and its specific asymmetry in the educational dialogue – is necessarily competent. In phenomenological and aesthesiological terms, being irreducible both to the five senses and to the perimeter of the skin, the felt body is therefore the main pedagogical tool through which things are learned, regardless of whether the process is mimetic or playful, ritualized or rationally finalized. The lived or felt body, which is normally repressed, does not so much provide an access to the world – a metaphor of mediation (which then fatally becomes interpretation, decodification or even construction) that is gravely misleading with respect to the fact that we are already always in-the-world in the form of perceptual presence (cf. Wiesing, 2014). Rather, it is the absolute place or zero point of our already being-in-the-world: and this, as I have already mentioned, is precisely thanks to its function as a faithful sounding board of the atmospheric feelings effused in our pericorporeal space.

I have already said that, willingly or unwillingly, the percipient always establishes a specific felt-bodily communication with every form of their world-environment, based on bridge-qualities such as motor suggestions and synaesthetic characters. Adults understand things cognitively first, in the hope that the body then will embrace the surrounding expressive-motor qualities and transform them into spontaneous behaviour. Instead, pathic inter-corporation is obviously more intense in subjects that are particularly receptive to such affective qualities, such as children and teenagers. In other words, children learn pathically what perceptively happens to them, well before becoming aware of it and rationalizing its content cognitively and linguistically: just as they learn how to swim, they also learn which emotional tonalities are positive and which are not in a felt-bodily way, and do so faster than adults.[5] And here the pathic interpretation needs once again an ethical integration. In fact, the undoubted human need for structures and rules, as a perpetual search for social tools aimed at reducing complexity, must be balanced in the learner by the satisfaction of the equally necessary need for freedom. The atmospheric involvement generated for example by charismatic personalities, who know how to exploit the well-known charm exercised on young people by the rhythmic suggestions inherent in any form of ritualization,[6] risks being absolute. To avoid this, it is necessary for the closeness to coexist with (and not exclude) distance (cf. Schultheis, 2013).

The point is obviously delicate, as it is difficult to establish when a happily pre-reflective participation in a powerfully atmospheric ritual turns into uncritical adherence to suggestive impressions aimed at manipulation – in short, at what point the educational pathicity becomes a pathological inhibition. The only possible answer (see Chapter 14) is twofold. On the one hand, the educator must responsibly avoid abusing the aesthetic-atmospheric element, for example by always showing the contingency and consequently also the problematic nature of what is being staged (cf. Schultheis, 2008, p. 112). On the other hand, the learner must be allowed to follow the immersive-empathic phase, which is essential to their training process, with a more sober critical-distancing phase, made possible by the school itself or, at least, by social life in its hopefully compensatory function. In the neo-phenomenological lexicon, this means that, in the educational felt-bodily communication, the almost ecstatic excorporation of the child (prevalence of wideness), produced by the predominance of encorporation in the charismatic educator (prevalence of tightness), must be followed by an encorporation also on the part of the child. Only by preserving their encorporation from the aggression caused by a unilateral charismatic encorporation (cf. Schmitz, 2013b) can the weaker subject escape the sometimes narcoticizing influence of an atmosphere.

But these risks should not push us to throw the baby out with the bathwater. Let's rather welcome the ingenious methods recently put in place to verify the type of atmosphere that is generated in a certain educational context, that is to say the degree of "aisthetic" (i.e. sensory in a broad sense) involvement of those exposed to it.[7] However, it must be clear that my atmospherological approach rejects the undoubtedly idealized Humboldtian and post-Enlightenment conception of education as a process of self-formation-cultivation, where the self is considered absolutely autonomous and therefore must be allowed to grow independently of any "operational" intent (even broadly understood). Also, my approach is opposed to the more recent but equally deleterious – if exclusive – attention to the practical, if not purely technological, aspects of education. The first trend is responsible for a somewhat plant-organic model of individual development; the second is the cause of a quantitative-teleological model that dissolves the most complex formative process into instrumentally planned actions. And both seem to seriously underestimate[8] the educational influence of atmospheres understood as emotionally tuned spaces.

Of course, in the name of traditional "whataboutism", some will object that the issues are more complex than that, that the school generates one or more atmospheres on the learners and the environment outside of school but at the same time it presupposes many other atmospheres of generically environmental origin (social, family, institutional atmospheres, etc.). But this cannot be an alibi to escape the phenomenologically evident fact that one learns mainly pathically, and that this circumstance deserves philosophical attention. And this attention, of course, is strongly conditioned: while the young are immediately exposed to the pedagogical atmosphere, not being able to describe or classify an emotional tone so fundamental as to be inadvertent, adults can only analyze it at distance. In so doing, adults make use of that ex-centricity (Helmut Plessner) that perhaps is not the most vibrant colour of our existence, but that is undoubtedly the blessing and curse of philosophy: even in its pedagogical variant, denying any residual intuitionistic illusion, philosophy can only be a reflection on the unreflected – in this case an atmospherological reflection on the relatively involuntary pedagogical atmosphere.

Notes

1 Which he himself somewhat modified in his latest works (cf. Schmitz, 2014, p. 75).
2 On the atmosphere of the gaze cf. Griffero (2017a, pp. 93–101).
3 For a synoptic view of this felt-bodily dynamic cf. Schmitz (2011).
4 Tellenbach (1968) notoriously focuses on the psychopathological consequences of a defective initial atmospherization.
5 They also learn this way how to distinguish good from evil, for example in a fairy tale, through the external appearance and the typified movement of its characters. See Schultheis (2008, pp. 103–104).
6 Which is often wrongly degraded to a negative remain of traditional authoritarian pedagogy.
7 According to Rauh (2012, pp. 221–223) there are three steps (obviously both useful and problematic) in this method: 1) synchronic annotation of perception, 2) reflective-anamnestic integration, 3) identification of the person collecting the data and the one that evaluates them. According to Wolf, instead, (2015, pp. 18–19), the steps are four: 1) observation of the situation, b) interrogation of the children's experience, 3) integration (especially for preschool children) of their verbal interviews with playful forms designed to give an account of the affective state experienced, and finally 4) asking them who and what in the room provokes in them this or that state of mind.
8 And wrongly so, given the undoubted incidence of the atmospheric on human growth and formation in the interaction with their *Umwelt*.

16 Theatrical atmospheres

Much ado about nothing?

We have thus come to the last of some (of the many possible) exercises of application of the atmospherological paradigm. This exercise comes last but not least: indeed, like architecture, theatre is a field of application that has always been privileged for the analysis of how atmospheric feelings are generated and perceived. And yet, until not long ago one could justifiably complain about the fact that "scholarly works devoted to theatre and the phenomenon of atmosphere are [...] surprisingly scarce" and therefore conclude that "the phenomenal relationship between atmospheric generation and the act of attending remains relatively unexplored" (Home-Cook, 2015, p. 197). Now, in the light of the often mentioned interdisciplinary "affective turn" in the humanities oriented towards qualitative studies, things have changed significantly. Indeed, the notion of "atmosphere", today so inflated, is perfectly at home in theatre studies as well as (or even more than) in all so-called aesthetic works. In short: wherever emphasis is put more on felt-bodily experience than meanings, more on emotionally arranging an environment than narratively representing something,[1] the atmospherological approach appears to be increasingly necessary.

Given that "the performative space always also creates an atmospheric space", one could say that atmospheric feelings "usually constitute the spectators' first sensation on entering the auditorium and enable a very specific experience of spatiality" (Fischer-Lichte, 2008, pp. 114–115). This is particularly true when the performances are able to breach the barriers of the audience's bodies and bring out its transformation. This does not imply, however, that theatre studies have yet been able to usefully apply the concept of atmosphere in a really new way. As said, one can certainly recognise an inherent relationship between atmosphere production and the act of staging, emphatically repeating that theatre is the readiest model for an aesthetics of atmosphere (Home-Cook, 2015, p. 173) and stressing the fact that, at least from baroque parties to Wagner's *Gesamtkunstwerk* and Reinhardt's non-naturalistic idea of interaction between stage and spectators (Schouten, 2011, pp. 178–191), theatre has always provided an essential trope of atmospheric multimodality (Curtin and Roesner, 2015, p. 112). However, this does not suffice to show the innovative value of the notion of atmosphere. The essential questions – To what extent does the attention move from the stage to the holistic theatrical space? To what extent does it generate a site-specific atmosphere? – so far have only been sketched, and they deserve a more detailed examination.

Without claiming to have any real competence in this field, but also without assuming any strict ontological distinction among everyday and artistic atmospheres (at least initially), I'll start by repeating that atmospheres are quasi-thingly, widely intersubjective, holistic feelings effused in space, i.e. prerifectively perceived intermodal affordances and authoritative generators of specific felt-bodily processes that precede (and/or deactivate) the traditional distinction between subject and object, between internal and external. Thus, atmospheres modulate the lived and pre-dimensional space in which we find ourselves, and therefore also a highly specific place such as the theatre hall. Now, even if an atmosphere lies not so much in the eye of the perceiver, but is rather a relatively objective feeling we encounter in the external lifewordly space – and theatre, of course, is an artform based on lived space and felt-bodily presentness more than many other media – I do not embrace *in toto* Hermann Schmitz's campaign for the desubjectification of feelings. With the purpose of a wider practical artistic and mediatic applicability of this approach, I prefer to admit, as often mentioned, that there are various types of atmospheres (prototypical, derivative and even quite spurious).

But one of the key points, possibly the "source of all problems" for any theory of atmospheres, deserves particular attention, because it concerns theatre perhaps more than other fields. I have repeatedly mentioned that, according to Schmitz, true atmospheres cannot be intentionally produced. The time has come to have a closer look at this question by taking advantage of the "case" of theatre. Let's start by saying, half-jokingly, that if Hermann Schmitz were fully right, then this analysis of theatrical atmospheres would really make little sense. If it were true, in fact, that there are theatrical atmospheres, but they cannot be produced since they come and go in a groundless way, it would be impossible to explain their genesis. One might well underline that in theatre – a word used necessarily in the broadest sense at this point – "nothing lives without atmosphere" (Brown, 2010, p. 143), but one ought then to explain how this special in-between-phenomenon can (also) be produced (with probable success). Let me be clear: this does not mean that we should underestimate Schmitz's critical remarks on the often banal instructions on how to produce atmospheric feelings, since, on the contrary, they could help in developing a better understanding of staging atmospheres. When he excludes as a matter of principle their producibility, however, he in good company and can count on a strong tradition. Let me give you a few but certainly leading examples.

According to Martin Heidegger (1995, pp. 88, 59, 65, 68), for example, a *Stimmung* (attunement or also atmosphere) is "a hybrid, partly objective, partly subjective", something we can neither invent ("least of all") nor "forcibly bring about, but into which we slip unaware", because it "must already be there". It is irreducible to intentional processes and cannot even be ascertained, since, if "all ascertainment means bringing to consciousness [...] all making conscious [an attunement] means destroying, altering in each case, whereas [...] we are concerned to let this attunement be as it is, as this attunement". It is worth adding that, even though one mostly feels them in their extreme manifestations, attunements and prototypical atmospheres are not "some being that appears in the soul as an experience": on the contrary, for Heidegger, "precisely those attunements to which we pay no heed at all [and] which attune us in such a way that we feel as though there is no attunement there at all, as though we were not attuned in any way at all – these attunements are the most powerful".

Following this approach, Otto Friedrich Bollnow, who must be given credit for publishing the first and most comprehensive study of *Stimmungen*, an arbitrary-intentional control over these attunements or moods is not an option because they are a unity existing before (in a neither spatial nor temporal sense) self and world and, therefore, also before any finalistic-conscious will. Even though human beings have a variable susceptibility to attunements, whose leading role goes largely unnoticed, they cannot free themselves from them but only contrast them with an opposite mood (obtained, for example, by fully devoting themselves to work) or by simply taking a positive or negative position on them, maybe thanks to some unexpected moral challenge (which forms the basis for a solid character). As such, an intentional production of moods, or mood-hygiene reached by detour, is exposed to fear of softness and would be "always the expression of a weakened life force linked to the serious danger of egotistically emptying the objective substance of life" (Bollnow, 1956, p. 140). Now, it's true that Heidegger and Bollnow talk more about moods or attunements (*Stimmungen*) than atmospheres, but it's clear that the two concepts are traditionally almost equivalent or barely discernible (Hasse, 2014, pp. 227–248). The most that can tentatively be said at this point is that, whereas atmospheres are external-spatial feelings, attunements are their subjective and spatially non-located tone (Schouten, 2011, p. 70).

Hermann Schmitz vigorously excludes that atmospheres may be produced: in previous works he provocatively sees them as something that lives outside and is objective almost like a road, certainly influenced here by Ludwig Klages' theory of daemons-images and his hard criticism of what spirit and will imply. He was later urged to take a position on this issue by Gernot Böhme, who indeed, while equally critical of the projectionist explanation of external feelings, admits that unlike the atmospheric – that is, something more objective that can be simply unaffectively established at a distance – atmospheres always imply the perceiver's felt-bodily participation (Böhme, 2001, p. 59 f.). Moreover, one is normally able to produce atmospheres and, however, to intersubjectively communicate about them. According to him, Schmitz's atmospheres "are free-floating like gods and as such have nothing to do with things, let alone being their product", so that "at most, objects can capture atmospheres, which then adhere to them like a nimbus".

Setting his aesthetics of production against Schmitz's aesthetics of (only) reception and stressing that, despite being experienced as something out there, "without the sentient subject, they [atmospheres] are nothing", Böhme underlines the "extraordinarily rich wealth of knowledge of atmospheres in the practical knowledge of aesthetic workers". In his view, thanks to their skills, aesthetic workers know well, although often tacitly, how and by what means to "create" "atmospheres by means of work on an object". This obviously explains the successful career of atmospheres in art and media studies: "atmospheres are involved wherever something is being staged, wherever design is a factor – which now means almost everywhere". This "creation" of atmospheres is somewhat only mitigated by the fact that the aesthetic-atmospheric competence is "confined to setting up the conditions in which the atmosphere appears" (Böhme, 2017a, pp. 17, 29, 20, 31), i.e. to establishing the generators through which, as conditions of appearance, atmospheric phenomena as tuned spaces "could" emerge (even as ecstasies of things (Böhme, 1995, pp. 199–200). Still, Schmitz's anti-intentionalistic concerns must be taken seriously and three main points deserve to be particularly highlighted.

1) When criticizing the assumption that one is able to "make" atmospheres and condense them into certain situations, maybe in order to escape from the solipsism of our societies through a constant "barrage of impressive, actual situations", Schmitz (1998, p. 182) is interpreting what Böhme defines "aesthetic work" exclusively in a negative way. As a technique of impression (*Eindruckstechnik*), according to his philosophy of history, it is nothing but the residue of the golden season of great European art. In other words, by manipulating "actual impressive situations" our age would only seek to finalistically stabilize them for propagandistic or advertising purposes (*Plakatsituation*): a strategy which unfortunately was brilliantly developed by Hitler, thus rhetorically giving birth, with nefarious consequences, to proto-aesthetic demands. The central problem would lie in the fact that intentionally produced atmospheres only aim at something apparent and superficial – to something that, indeed, one can only "sense"(!).

2) From a more ontological point of view, Schmitz claims that accepting the intentional generability of atmospheres through the skilful utilisation of ecstasies of things means accepting the singularistic false assumption that situations are made up primarily of single objects, whose constellation would possibly later elicit an atmospheric effect, and are not, therefore, a holistic, i.e. chaotic-multiple character. At best, single things could be the quasi-thingly medium of more widespread atmospheres, but certainly not their formative source. In their original (i.e. prototypical, in my terms) form, they are so dynamic that – as Schmitz and I never tire of saying (Schmitz, 1998, p. 188; Griffero, 2014a, pp. 119–120) – they come and go, without there being any point in asking what they did in the meantime. 3) Lastly, by stating that an atmosphere is radiated by a thing, or even a city, one would uncritically adopt the mistaken dualistic sender-receiver model (Schmitz, 2003, p. 244) and thus fail to appreciate that an atmosphere is rather an enveloping pre-dualistic

lived space, something that as such appears even without a motivating background.

This apparent digression makes it abundantly clear that the contrast between these two paradigmatic positions has wider ontological-philosophical implications. It also refers, however, to a different assessment of today's aestheticization of everyday life, stigmatized by Schmitz as a misleading propaganda and welcomed instead by Böhme as the (neutral) unavoidable reflection of the "theatricalization of life" (Böhme, 2017a, p. 33) typical of aesthetic capitalism (Böhme, 2017c). It is no coincidence that Böhme regards the same term rejected with disdain by Schmitz – "technique of impression" – very appropriate and apt (Böhme, 2017a, p. 33) for expressing the widespread and positive (or at least neutral) fact of bringing something to appearance. There are many possible replies to Schmitz's point. A) The first thing to say is that his stigmatization conflicts with Schmitz's idea that dwelling, church, gardens and Japanese tea houses would perfectly capture and positively cultivate atmospheric feelings in a closed space, taking for granted – who knows why! – that dwelling as a culture of feelings within a self-contained space can be non-manipulative. B) Secondly, Schmitz wrongly downgrades the rhetorical *coté* of the aesthetic tradition to something that only generates superficiality and precariousness, thus depriving it of its educational and cultural value. C) Thirdly, assuming that politics, media and advertising impressions are only the ingredients of fake atmospheres implies an axiology which is seriously problematic within a context that claims to be rigorously descriptive-phenomenological, especially because only external and non-phenomenological parameters would allow us to distinguish between good (true) and bad (false) feelings.

My point of view, in a certain sense, poses fewer problems and stems from my analysis of the origin of atmospheric power. My "inflationary" distinction among three kinds of atmosphere allows me to not set Schmitz's position – according to which the only authentic atmospheres are those that have, so to speak, a transcendent reality and are thus totally independent of humans and things – radically against that of those authors, like Böhme, who try instead to mitigate the overly random nature of atmospheres and to bind them as strongly as possible to objectual poles of which they would be the ecstasies. This doesn't mean, however, that all atmospheres derive from things (or actions). Indeed, space itself (without things) can be felt as the real medium of an atmosphere (Hasse, 2012, pp. 25–26). Prototypical atmospheres – which in this way perhaps partly overlap with *Stimmungen* – hover around us without being strictly linked to causes and objects and without being intentionally produced. However, one should honestly admit that all aesthetic works are perfectly able to produce at least the second type of atmosphere, namely the derivative one, understood as an in-between that sometimes even acts more as a condition of possibility than as an outcome of the subject-environment relationship. But now one should ask to what extent an atmosphere can indeed be planned.

What does it mean, exactly, to design an atmosphere? In so doing, do we vaguely feel it or just imagine it? Shouldn't we perhaps resign ourselves to the fact that, because of their unavoidable vagueness, openness and random pattern, they are something unavailable and that, for this reason, a potential atmosphere is never an atmosphere in the full meaning of the word? There are no recipes, of course, for staging atmospheres, and their producibility varies largely depending on the different fields. The designed atmosphere of a landscape garden, for example, depends on unplanned events much more than a (traditional) stage production. Exactly like architectural drawings (cf. Chapter 9), all one can do by staging a play is *try* to express its desired atmosphere by various means and as accurately as possible. Like the designer and the urbanist, thanks to their prognostic body- (or motor)-schema, the theatre director should stimulate a developed role-play competence and some form of complicity with the audience, predicting the latter's future felt-bodily resonance and identifying as far as possible with its affective-cognitive situation. However, as these are nothing more than counterfactual hypotheses about an area that they obviously cannot fully control, the director can only try to design certain effects, but without ever being able to escape the heterogenesis of ends that, threatening every human activity, can never be completely avoided in theatre either.

After all, if an atmosphere only existed when clearly subjectively felt, could we really carry out an abstract analysis of the conditions of possibility of its production? Could we intersubjectively talk about it (thereby somehow altering and falsifying the effect felt before)? The question is theoretically legitimate, of course, but I would suggest not to overestimate the strict actuality of atmospheres. First of all because they undoubtedly always depend also on the co-perception of other past and expected atmospheres, such as when, for example, you feel the tense atmosphere of a play precisely because you anticipate the following situation, heavily influenced by title, name of the author, genre, previous performances, etc. Secondly, because the actual atmosphere may also be the successful outcome of a counterfactual staging, in the sense that a director can easily imagine under what well-designed conditions the atmosphere of that play can be more or less tense, more or less relaxing, in short even a success or a failure: the endless rehearsals before the event, even in the case of unconventional performances, aim to ensure this emotional effect (cf. Schouten, 2014, p. 139). As a result, although a planned atmosphere obviously only exists in an incomplete and potential way, I don't think it's true that theatre, like other aesthetic works, never actually produces atmospheres but only evokes them in the spectator in a vague and imprecise way. Arranging spatio-temporal scenes that successfully invite certain viewers to know and feel something implies being perfectly aware of how to design theatrical affordances-atmospheres and how they are perceived by an audience that, precisely because it is captured by them, ends up feeling completely involved by the play.

Let's now take a closer look at theatre, while being fully conscious that, since I'm tiptoeing through a minefield, my inexperience in this field might play some tricks here. I see no need to repeat in this specialized context that theatre is a place professionally assigned to producing an atmospheric situation less volatile than the everyday *mise-en-scène* that we define theatricality (cf. Rodatz, 2010, pp. 101–109), or that theatre counts on that "unique phenomenon of a distance, however close it may be" that Benjamin defines "aura". Nor is it necessary to decide whether theatrical atmospheres are due more to the light's ability to create a specific spatial attunement (cf. Kümmerlen, 1929, 42 f.) or rather to the enveloping sound, including even a voluminous silence (cf. Ströker, 1977, p. 28 and Hauskeller, 1995, p. 53). Rather, it is important to know where and how light and sound, more specifically, cease to serve as diegetic background phenomena or as signs referring to something other, i.e. as by-products of a performance still designed for a sign-based perception, and develop instead independent feelings designed for an atmospheric perception.

An atmospherological approach to theatre will obviously welcome the increasing attention to what emotionally and intermodally (and even synaesthetically) fills the space "between and amongst actors and audience" (Welton, 2012, p. 130), also embracing the emphasis on the many generators or "atmospherical amplifiers" (Schouten, 2014, p. 140) of this specific lived space. This includes "the shift in temperature between inside and outside, and between what is represented and what is experienced", well knowing that, as in any other field, "natural and 'man-made' atmospheres co-mingle" (Welton, 2012, pp. 138, 141) also in theatre. But if an atmospherological approach were limited to these points, it would have undoubtedly been anticipated and fulfilled by a talented forerunner like Mikhail Chekhov. In his view, an atmosphere is a visual and sensory text based on all that can be perceived on stage and that "energetically" connects the people on stage to the people in the audience (and actors and audience to each other).

As a sort of tension "spread in the air, enveloping people and events, filling the rooms, floating through the landscapes, pervading the life of which it is a part" (Chekhov, 2002, p. 54), the theatrical atmosphere, while changing based on scenes, dialogues, etc. and depending on the intervention of the public, would be the objective feeling that fights against, and necessarily prevails over, all feelings taking place throughout the play's "score". Moreover, Chekhov pays special attention to conflicting atmospheres of different scenes and to discrepancies between the atmosphere of the scene and the individual atmosphere of each character. Thus, the atmosphere would be the means for a double-framing (cf. Meerzon, 2005, p. 272) which gives life to a strong (and circular) bond between actors and (co-experiencing) spectators and, at the same time, "deepens the perception of the spectator" (Chekhov, 2002, p. 48) in such a way that actor and spectator become simultaneously both sender and receiver. Despite his originality, Chekhov is still too subordinate to a "signaletic" and semiotic conception of stage-audience interactivity, being thus forced to traditionally distinguish between subject and object and to refer to extra-theatrical realities, whose atmospheric function only finds its *conditio sine qua non* within the framework of the overall atmosphere of the performance (cf. Schouten, 2011, p. 110).

In any case, a more up-to-date approach to theatrical atmosphere must take into account the major epochal change that took place in the 20th century, that is, the advent of postdramatic theatre. Understood as "a trend for performances which use installations and expansive environments, which have mobile audiences, and which invite audience participation" (White, 2012, p. 221), this theatre gives up the psychological-realistic representation. Now, traditional illusion theatre, initially theorized by Lessing and eventually killed by Brecht, could already be considered a performative act which, for this reason, counted on various atmospheric effects usually based on the audience's identification with characters and plot. Therefore, there can be no doubt that postdramatic theatre, in the light of the so-called performative turn, must pay special attention to atmospheric means. The theatregoer, who today could even be defined as a "guest", insofar as they have an aesthetic experience that is exclusively possible in this type of theatre (cf. Rodatz, 2010, p. 151), generates a specific atmosphere through all their conducts and behaviours, thus contributing to the scenic one. The audience especially focuses their attention on their own emotional perception rather than being narratively redirected to extra-theatrical realities. This peculiar perceiver-perceived reciprocity, elicited by the challenge of every traditional correspondence-based system (among theatrical signs, characters and their social status or role in that play, etc., as well as between the performance and its reception) is precisely what postdramatic atmosphere means (cf. Schrödl, 2009).

This performative turn has made theatre more a "presence" than a representation, more an intersensorially shared experience than a unilateral communication, more a process than a product, more a manifestation than a signification, generally speaking more an energetic impulse than information (cf. Lehmann, 2001, p. 85). Thanks to its cross-cultural atmospheres, this non-conventional approach, by no longer using voices, images, sounds and smells[2] as media of psychological characters, becomes understandable to spectators anywhere and demands being considered as a form of immersion which, without ever fully neglecting a semiotic and hermeneutic approach, substantially changes the hierarchy of scientific points of view. Now, regardless the validity of an inflated slogan like "immersion", whose meaning as a temporal cessation and a radical seizure of conscience (cf. Home-Cook, 2015, p. 140), as already noted, could be appropriate only to real (and not fictional) perception, I should point out that many reflections on postdramatic theatre often run the risk of considering really atmospheric what are in fact quite exceptional experiences such as the presence of something unfamiliar and indecipherable, if not threatening. Too influenced by the avant-garde ideals of shock and fusion between art and life, many reflections assume that a theatrical atmosphere only emerges when there is no attunement between stage and viewers, or when the intra-scenic tension also applies to the public (cf. e.g. Lehmann, 1999, p. 22 and Rodatz, 2010, pp. 59–60), when the line between stage and hall disappears, or when the audience fully participates in the event. The main negative effect of this exceptionalism is to ignore that there are also many (more conventional and auratic) theatrical atmospheres, which radiate into the lived space and tinge the in-between as much as the provocative ones, without denying illusionistic models.

But it is now time to address some important and controversial issues suggested by my atmospherological and neo-phenomenological perspective.

a) It should be first noted that the neo-phenomenological redefinition of perception as a felt-bodily communication with all that is perceived could provide a better explanation of the theatrical experience. The theatrical atmosphere, in fact, precisely expresses the "here and now" of a lived space and the way I feel situated. More precisely, it expresses both one's presence in space and the way one feels within it, even if, of course, on a channelling and controlled basis. During the performance, the felt-bodily communication between everything on the lit-up stage and everything in the darkened hall constantly generates intercorporeal and superordinate units in the form of encorporation or excorporation. Hence a wide range of possible states, some positive (prevailing expansion) and some negative (prevailing contraction), according to the intertwining between the two extreme poles of the vital dynamic, namely "tightness" (contraction) and "wideness" (expansion) (*Enge/Weite*). All this, which is only mentioned here, should be studied in detail also from a theatrical standpoint.

b) Secondly, one has to acknowledge that atmospheres are not composed but "generated", namely that they supervene on their (material and immaterial) generators. Even those who do not believe in their systematic analysis have to recognise, in fact, that without phenomenal elements composing the designed theatrical environment, theatrical atmospheres would not exist either. Script, actors, sound, scenography, set design, lighting, the theatre's building and entrance, and even silence on stage[3] provide the initial stimulus for the manifestation of an atmosphere, first of all through the ecstasies by which the single components express themselves. Lacking a more developed theory of theatrical atmospheric supervenience, it could at least be said that these generators only act effectively if, as required by the ingressive paradigm and by modern theatre criteria, the performance succeeds from the first impression in spatially separating the theatregoers from their everyday life. This emotional change happens well before the performance and is already made possible by the entrance, the magnificence of the building, the darkened and quiet hall, the seating position, the light on stage focusing the viewer's attention on a limited place, etc. Thanks to this very heterotopic constraint (cf. Rodatz, 2010, p. 83 ff.) and its long-lasting rituality – we often enjoy even postdramatic performances in traditional theatre buildings, hence a very peculiar atmosphere of interference! – the mentioned atmospheric generators actually start up, thus giving life to a real theatrical "aesthetic difference" (to use Gadamer's expression). This initial emotional involvement therefore owes almost everything to a double cut through space: between indoors and outdoors as well as between stage and auditorium. Though based on factors such as duration, *mise-en-scène*, immunity from action, distinction between actual reality and material reality, differentiation between I-pole and object-pole, ingression and discrepancy (ibid., p. 141 ff.), the theatrical atmosphere is primarily due, however, to the fact that theatre, unlike other art forms, forces senders and receivers – to use these inappropriate terms – to spend together a period of their lifetime and to breathe the atmospheric "air" of the same space (cf. especially Lehmann, 2001, p. 12 f.).

c) Given that every atmosphere, including the theatrical one, properly exists in the full sense of the term, as repeatedly stated, only through the felt-bodily communication between stage and theatregoers, it is also legitimate to suppose that it always both shapes, and is shaped by, the intersubjective attentional enactments of the audience. The role of attending, let alone of attention, should nevertheless not be overestimated, even if attention were reconsidered in terms of "stretching" and not as a "spotlight". It's totally plausible, in fact, that the most influential atmospheres are exactly those which, as intermediate phenomena,[4] always reside on the margins of our attentional sphere. Preceding the rigid subject/object (here play/spectator) distinction, derivative theatrical atmospheres are sometimes imperceptibly determined by the inexpressible felt-bodily co-presence (or "feeling") of staying together in a particular space, whose timbre might also momentarily not be consciously noticed.

In other words, by their nature, these atmospheres tend to lie in the background, so it appears controvertible to suggest that they could only fully come into being as a consequence of their being noticed.[5] However, when by attention one means something more and different, that is, an ongoing, intersensorial, bodily engagement with the affordances of any given environment or atmosphere, or even a sort of "dynamic dance" (Home-Cook, 2015, p. 168), there actually is no reason to carry on talking about attention instead of the already mentioned felt-bodily communication. And yet, the latter goes largely unnoticed, without attracting attention at all. In my opinion, if the "how" of our attending shapes the "what" of our perceiving, this enactive "how" must still depend more on stage affordances than on the autonomous spectator's initiative. The aesthetics of reception would face serious difficulties in demonstrating that the audience is the generator of the tuned space and that, consequently, the atmospheric effect is "almost exclusively placed in the hands of the audience" (ibid., pp. 161–162).

d) It's undeniable, indeed, that an atmospheric experience is often based on a perceptive non-transitivity (non-theticity). Rather than perceiving something (even the atmosphere itself), one here would perceive "in accordance" with something, namely in accordance with an outside atmosphere. It follows that atmospheres are not so much percepts but rather non-transitive contextual conditions of a perception that only becomes transitive later. It can therefore be assumed that also in theatre they act as a horizon which, while affecting everything it lingers on, stays unobserved, much like "glasses, which cannot be seen when one looks through them" (Bockemühl, 2002, p. 221).[6] Theatre is thus a transcendental unnoticed sensory-perceptive ambiance specifying through a motor stimulation the conditions under which phenomena emerge and appear to us. Exercising rather indifferently in the "focalization, defocalization and refocalization" (De Marinis, 1987, p. 106) of the audience's attention, it probably exerts its greater influence in its defocalizing effects. On the other hand, even supposing that the full focus on the scene normally required in theatre may risks breaking the atmosphere, it is important to always ensure that the reduction of atmospheres to purely contextual qualities does not result in underestimating their quasi-thingly authority and significance. The best thing one can say is that their more situational or more objectual nature will depend most likely, case by case, on the kind of the performance in question, for example on its belonging to dramatic or post-dramatic theatre.

e) Lastly, common experience shows that a spectator sometimes perceives an atmosphere without being overwhelmed by it. This means that one lives situations in which neither the traditional immunity from practical action nor personal and bodily destabilization necessarily elicit deeply involving atmospheres. A performance that goes beyond the boundaries of distancing senses such as sight and hearing, and maybe violates the conventional primacy of actuality over reality, might not really capture the spectator, even though it would force them to dismiss their contemplative and distanced position through a provocative, unexpected and bodily involvement in reality, and thereby to find themselves face-to-face with physical reality. Despite not being problematic in its ontological version, the implied distinction between feeling and reflecting could therefore be much more elusive in theatre, where atmospheres clearly emanate from both expressive-affective qualities and mental processes, and for this reason need to be both semiotically and atmospherically understood.[7] Accordingly, the already mentioned "affective turn" does not necessarily involve a guilty blindness toward the unavoidable overlappings between emotion and cognition. The point, however, is whether a bodily involving performance like the postdramatic one, stimulating the spectator to be on constant standby without ever relaxing, can be really considered "atmospheric" in the "true" sense of the term.

As indicated earlier, for an oversimplified application of the atmospherological approach to different contexts one should distinguish between three types of atmosphere, which in turn, depending on their phenomenological encounter with the spectator, are then able to generate several types of emotional games as a variation of the in-between. Trying to briefly adapt this phenomenological differentiation to theatre as far as possible,[8] one should at least admit the following options. When an introductory phase leads the spectator into the desired mood through various (also semiotic) staging strategies, an atmosphere can overwhelm them (ingressive encounter) and be refractory to a more or less conscious attempt at a projective reinterpretation. When they are involved by the performance, therefore, the theatregoers still experience themselves, initially at least, as something other than the atmosphere encountered. After this first ingressive phase, the discrepant spatial feeling can elicit two different affective situations. One option is that it can invade the spectator and change their mood, now felt as contrasting with that one suggested by the performance as well as perhaps with the mood of other viewers. This happens especially if the "as if" narrative of the plot gives way to the felt-bodily (and bodily *tout court*) presentness of what is actually happening on stage. Although they cognitively feel something antagonistic, the spectator can, however, be touched by it only at a distance and without a real personal involvement, sometimes because their tension might be mitigated by the comforting presence of other spectators.

The second option is that an atmosphere can find the spectator in tune with it (syntonic encounter), to the point that they don't even realize they have entered it. But this situation is almost impossible in theatrical experience, which in any case implies, in fact, a separation from everyday life. It could happen, instead, that the performance's atmosphere does not meet the necessary threshold for a sensorial-affective participation, thus causing in the audience a full misunderstanding of the performance. Also, a spectator's excessively strong mood may prevent them from letting themselves go to what they perceive and sense during the performance. Furthermore, it often happens that one is able to clearly recognize an atmosphere and even to describe it well to others but without being really felt-bodily involved by it. This happens when a performance, while radiating a mood that is easily "readable" from a semiotic, cultural and physiognomic point of view, does not really capture the spectator. It may not be excluded that the failure of this distanced approach, for example following the inadequacy of a coherent semantic interpretation of what appears on stage, can later even elicit an unexpected atmospheric mood impossible until that time. The atmosphere can even generate resistance: when this happens in everyday life, the perceiver can be pushed to change it. In a theatrical situation, though, at most the spectator can leave the building unsatisfied or, when they cannot even tolerate what they are watching, they may vigorously and loudly protest against the performance.

For various reasons (cultural, social, political, aesthetical and even absolutely idiosyncratic), an atmosphere may of course also be perceived differently, especially in the course of time. In theatre, as well as in any art based on time, one can, in fact, initially feel something, but also change this mood during the performance, not least because, for example, they perceive contradictory details, become aware of errors or omissions in the script, notice the director's bad choices, etc. The atmosphere may be even so dependent on the perceptual (subjective) form as to concretize itself, both for the performers and for the viewers, even in "ingredients" that normally express completely different moods: it is what normally happens when avant-garde theatre overturns conventions and aims at shocking its audience through totally unexpected gestures and ecstasies of things. An atmosphere, finally, might elicit an active feedback loop to the extent that spectators, accepting the performance, coordinate with it to the point of reinforcing the staged atmosphere. They thus influence the actors and the other spectators, or even lead to atmospheres conflicting with the designed one. Instead, it is not so sure that, in cases like this one, the stronger mood necessarily eliminates the weaker one, as Chekhov instead claims, evidently underestimating that in many cases a performance also can end up suggesting a totally unresolved atmospheric oscillation without being a failure for this reason.

It goes without saying that this phenomenological classification, which applies to theatre some ideas I have often restated, while not being exhaustive on the one hand, could seem too abstract on the other. The truth is that a theatrical performance almost always arouses a mixed atmosphere not strictly limited to the options mentioned. This is clearly reflected in the fact that when one, paradoxically trying to do something impossible, verbalizes an atmosphere, one is forced to always add some elusive "somewhats" or "somehows". However, one must fully be aware of the essential vagueness of atmospherology and fruitfully capitalize this only apparent defect. It should especially be borne in mind that the atmospherological approach, primarily when it's not based on the ontological externalization of the affective, risks simply expressing the same thing (feelings, emotions) with different words (atmospheres) – a risk that is always inherent in every real or presumed paradigm shift. Despite this well justified doubt, I believe that judging this approach to theatre as a "much ado about nothing" would be a clear sign of small-minded thinking.

Whenever atmospheric feelings can be staged, scholars and researchers should engage in explaining how to make good use of feelings. In my opinion this always means two things: that every person should deeply and pathically feel atmospheres but also responsibly acquire the competence needed to critically evaluate them. One should never underestimate the fact that the atmospherological approach, when adapted to different fields of experience – as we have done here by applying it to some of the many more possible fields (architecture, dwelling, inclusive space, the *genius loci*, the earth as expressive quality, democracy, pedagogy and, finally, theatre) – is always declined a little differently and is forced to deal with specific and even discordant aspects that the general theory could not foresee. But this is a good thing, otherwise nothing would be learned from experience, whether in confirmation or denial. After all, to paraphrase a famous Tolstoyan *incipit*, one could say that "all atmospheres are (theoretically) alike, but each (concrete) atmosphere is atmospheric in its own way".

Notes

1 Unless, obviously, one gives an unusual and much wider definition to the term "narrative".

2 Especially the last two according to Fischer-Lichte (2008, pp. 116–118).

3 According to Home-Cook (2015), theatrical atmospheres most acutely draw our attention in moments of silence or hiatus.
4 In two ways: between environment and perceiver but also among all senses.
5 For this argument see Home-Cook (2015, pp. 156, 161).
6 An argument often used by Thibaud (2015).
7 See Schouten (2011, pp. 79, 81) and Rodatz (2010, p. 39), who here follows Böhme.
8 For a simpler overview, see Schouten (2011, p. 208 ff.).

Bibliography

Adorno, T. W. (1997). *Aesthetic Theory*. London and New York: Continuum.

Adorno, T. W. (2004). *Negative Dialectics*. London and New York: Routledge.

Adorno, T. W. (2005). *Minima Moralia: Reflections on a Damaged Life*. London and New York: Verso.

Alexander, C., Ishikawa, S. and Silverstein, M. (1977). *A Pattern Language*. New York: Oxford University Press.

Andermann, K. (2007). *Spielräume der Erfahrung. Kritik der transzendentalen Konstitution bei Merleau-Ponty, Deleuze und Schmitz*. München: Fink.

Andermann, K. and Eberlein, U. (eds). (2011). *Gefühle als Atmosphären. Neue Phänomenologie und philosophische Emotionstheorie*. Berlin: Akademie Verlag.

Anders, G. (1956). The World as Phantom and as Matrix. *Dissent* 3, 1, pp. 14–24.

Arnheim, R. (1966). *Toward a Psychology of Art*. Berkeley, CA: University of California Press.

Arnheim, R. (1974). *Art and Visual Perception: A Psychology of the Creative Eye*. Berkeley, CA and London: University of California Press.

Augoyard, J.-F. (2016). L'apparaître des ambiances. In Rémy, N. and Tixier, N. (eds), *Ambiances, Tomorrow*, International Ambiances Network – University of Thessaly, II, pp. 579–585.

Avenarius, R. (2014). *Der menschliche Weltbegriff. Mit einer Einführung von Hermann Schmitz*. Berlin: Xenomoi.

Bachelard, G. (1948). *La terre et les rêveries d la volonté*. Paris: José Corti.

Bachelard, G. (1994). *The Poetics of Space*. Boston, MA: Beacon Press.

Baensch, O. (1923). Kunst und Gefühl. *Logos* XII, 1, pp. 1–28.

Baudelaire, C. (1996). *Artificial Paradises*. Bethesda, MD: Carol Publishing Group.

Baudrillard, J. (2002). *The System of Objects*. London and New York: Verso.

Bausinger, H. (1988). Räumliche Orientierung. Vorläufige Anmerkungen zu einer vernachlässigten kulturellen Dimension. In Bringéus, N.-A. (ed.), *Wandel der Volkskultur in Europa. Festschrift f. G. Wiegelmann*. Münster: Coppenrath, I, pp. 43–52.

Benjamin, W. (2002). *The Arcades Project*. Harvard, MA: Harvard University Press.

Bernet, R. (2014). Das Subjekt des Leidens. In Fuchs, T., Breyer, T., Micali, S. and Wandruszka, B. (eds), *Das leidende Subjekt. Phänomenologie als Wissenschaft der Psyche*. Freiburg-München: Alber, pp. 11–32.

Bieger, L. (2011). Ästhetik der Immersion: Wenn Räume wollen. Immersive Erleben als Raumerleben. In Lehnert, G. (ed.), *Raum und Gefühl. Der Spatial Turn und die neue Emotionsforschung*. Bielefeld: [transcript], pp. 75–95.

Bischoff, W. (2005). "Grenzenlose Räume" – Überlegungen zum Verhältnis von Architektur und städtischem Geruchsraum. *Wolkenkuckucksheim* X, 2. www.cloud-cuckoo.net/openarchive/wolke/deu/Themen/052/Bischoff/bischoff.htm.

Blume, A. (2003). *Scham und Selbstbewusstsein. Zur Phänomenologie konkreter Subjektivität bei Hermann Schmitz*. Freiburg-München: Alber.

Blume, A. and Demmerling, C. (2007). Gefühle als Atmosphären. Zur Gefühlstheorie von Hermann Schmitz. In Landweer, H. (ed.), *Gefühle – Struktur und Funktion*. Berlin: Akademie Verlag, pp. 113–133.

Bockemühl, M. (2002). Atmosphären sehen. Ästhetische Wahrnehmung als Praxis. In Mahayni, Z. (ed.), *Neue Ästhetik. Das Atmosphärische und die Kunst*. München: Fink, pp. 203–222.

Böhme, G. (1989). *Für eine ökologische Naturästhetik*. Frankfurt a. M.: Suhrkamp.

Böhme, G. (1994). *Weltweisheit-Lebensform-Wissenschaft. Eine Einführung in die Philosophie*. Frankfurt a. M.: Suhrkamp.

Böhme, G. (1995). *Atmosphäre. Essays zur neuen Ästhetik*. Frankfurt a. M.: Suhrkamp.

Böhme, G. (1998). *Anmutungen. Über das Atmosphärische*. Ostfildern v. Stuttgart: edition tertium.

Böhme, G. (2001). *Aisthetik: Vorlesungen über Ästhetik als allgemeine Wahrnehmungslehre*. München: Fink.

Böhme, G. (2002). *Die Natur vor uns. Naturphilosophie in pragmatischer Hinsicht*. Kusterdingen: Die Graue Edition.

Böhme, G. (2006). *Architektur und Atmosphäre*. München: Fink.

Böhme, G. (2008). *Ethik leiblicher Existenz*. Frankfurt a. M.: Suhrkamp.

Böhme, G. (2010). *Anthropologie in pragmatischer Hinsicht*. Bielefel-Basel: Sirius.

Böhme, G. (2012). *Ich-Selbst. Über die Formation des Subjekts*. München: Fink.

Böhme, G. (2013). Wirklichkeiten. Über die Hybridisierungvon Räumen und die Erfahrung von Immersion. *Jahrbuch Immersiver Medien*, pp. 17–22.

Böhme, G. (2014). *Bewusstseinsformen*. München: Fink.

Böhme, G. (2017a). *The Aesthetics of Atmospheres*. London and New York: Routledge.

Böhme, G. (2017b). *Atmospheric Architectures: The Aesthetics of Felt Spaces*. London: Bloomsbury.

Böhme, G. (2017c). *Critique of Aesthetic Capitalism*. Milan: Mimesis International.

Böhme, G. and Böhme, H. (1985). *Das Andere der Vernunft: Zur Entwicklung von Rationalitätsstrukturen am Beispiel Kants*. Frankfurt a. M.: Suhrkamp.

Böhme, G. and Böhme, H. (2004). *Feuer, Wasser, Erde, Luft. Eine Kulturgeschichte der Elemente*. München: Beck.

Böhme, H. (1989). Das Steinerne. Anmerkungen zur Theorie des Erhabenen aus dem Blick des 'Menschenfremdesten'. In Pries, C. (ed.), *Das Erhabene. Zwischen Grenzerfahrung und Größenwahn*. Weinheim: Acta Humaniora, pp. 160–192.

Böhme, H. (1992). Gaia: views on earth from Hesiod to James Lovelock. In Niethammer, L. (ed.), *Bericht 1991 des Kulturwissenschaftlichen Instituts*. Essen: Wissenschaftszentrum Nordrhein-Westfalen, pp. 195–211.

Bollnow, O. F. (1956). *Das Wesen der Stimmungen*. Frankfurt a. M.: Klostermann.

Bollnow, O. F. (2001). *Die pädagogische Atmosphäre. Untersuchungen über die gefühlsmäßigen zwischenmenschlichen Voraussetzungen der Erziehung*. Essen: Die Blaue Eule.

Bollnow, O. F. (2011). *Human Space*. London: Hyphen Press.

Borch, C. (2011). Foamy business: On the organizational politics of atmospheres. In Schinkel, W. and Noordegraaf-Eelens, L. (eds), *In Medias Res. Peter Sloterdijk's Spherological Poetics of Being*. Amsterdam: Amsterdam University Press, pp. 29–42.

Borch, C. (ed.) (2014). *Architectural Atmospheres: On the Experience and Politics of Architecture*. Basel: Birkhäuser.

Bourdieu, P. (2010). *Distinction: A Social Critique of the Judgement of Taste*. London and New York: Routledge.

Bozzi, P. (1998). *Fisica ingenua. Studi di psicologia della percezione*. Milano: Garzanti.

Bronner, G. (2009). *La pensée extrême. Comment des hommes ordinaires deviennent des fanatiques*. Paris: Denoël.

Brönnle, S. (2009). Geomantie als Brücke zwischen den geistigen Ebenen von Mensch und Ort. In Mallien, L. and Heimrath, J. (eds), *Genius Loci. Der Geist von Orten und Landschaften in Geomantie und Architektur*. Klein Jasedow: Drachenverlag, pp. 51–60.

Brown, R. (2010). *Sound: A Reader in Theatre Practice*. Basingstoke, UK: Palgrave Macmillan.

Bühler, K. (1933). *Ausdruckstheorie. Das System an der Geschichte aufgezeigt*. Jena: Gustav Fischer.

Bulka, T. (2015). *Stimmung, Emotion, Atmosphäre. Phänomenologische Untersuchungen zur Struktur der menschlichen Affektivität*. Münster: mentis.

Bürklin, T. (2008). Sitte "reloaded". Wohnen in der Stadt, die es nicht mehr gibt. In Hasse, J. (ed.), *Die Stadt als Wohnraum*. Freiburg-München: Alber, pp. 149–167.

Busch, K. (2007). Ansteckung und Widerfahrnis. Für eine Ästhetik des Pathischen. In Busch, K. and Därmann, I. (eds), *»pathos«. Konturen eines kulturwissenschaftlichen Grundbegriffs*. Bielefeld: [transcript], pp. 51–73.

Casey, E. S. (1993). *Getting Back into Place: Toward a Renewed Understanding of the Place-world*. Bloomington, IN: Indiana University Press.

Chekhov, M. (2002). *To the Actor: On the Technique of Acting*. London and New York: Routledge.

Claus, L. F. (1926). *Rasse und Seele. Eine Einführung in die Gegenwart*. München: Lehmann.

Cresswell, T. (2004). *Place: A Short Introduction*. Hoboken, NJ: Wiley-Blackwell.

Curtin, A. and Roesner, D. (2015). Sounding out 'the Scenographic Turn': Eight position statements. *Theatre & Performance Design* 1, 1–2, pp. 107–125.

Damisch, H. (2001). *Skyline: The Narcissistic City*. Stanford, CA: Stanford University Press.

De Botton, A. (2006). *The Architecture of Happiness*. New York: Vintage Books.

De Marinis, M. (1987). Dramaturgy of the spectator. *Drama Review* 31, 1, pp. 100–114.

Demmerling, C. (2011). Gefühle, Sprache und Intersubjektivität. Überlegungen zum Atmosphärenbegriff der neuen Phänomenologie. In Andermann, K. and Eberlein, U. (eds), *Gefühle als Atmosphären. Neue Phänomenologie und philosophische Emotionstheorie*. Berlin: Akademie Verlag, pp. 43–55.

Demmerling, C. and Landweer, H. (2007). Philosophie der Gefühle – Ansätze, Probleme, Perspektiven. Zur Einleitung. In Demmerling, C. and Landweer, H. (eds), *Philosophie der Gefühle. Von Achtung bis Zorn*. Stuttgart: Metzler, pp. 1–34.

Derrida, J. (1991). *Of Spirit: Heidegger and the Question*. Chicago, IL: University of Chicago Press.

De Sousa, R. (1987). *The Rationality of Emotion*. Cambridge, MA: MIT Press.

Dewey, J. (1931). *Philosophy and Civilization*. New York: Minton, Balch & Company.

Döring, S. A. (2011). *Gründe und Gefühle: Zur Lösung "des" Problems der Moral*. Berlin and New York: de Gruyter.

Dürckheim, K. G.von (2005). Untersuchungen zum gelebten Raum. Erlebniswirklichkeit und ihr Verständnis. Systematische Untersuchungen II. In Hasse, J. (ed.), *Graf Karlfried von Dürckheim Untersuchungen zum gelebten Raum*, mit Einführungen J. von Hasse, A. Janson, H. Schmitz and K. Schultheis, Frankfurt a. M.: Selbstverlag Institut für Didaktik der Geographie, pp. 11–108.

Ellgring, H. (2010). Mimischer Ausdruck von Emotionen – Assoziation und Dissoziation von Erleben und Verhalten. In Grossheim, M. and Volke, S. (eds), *Gefühl, Geste, Gesicht. Zur Phänomenologie des Ausdrucks*. Freiburg-München: Alber, pp. 56–87.

Falter, R. (2003). *Ludwig Klages. Lebensphilosophie als Zivilisationskritik*. München: Telesma.

Falter, R. (2006). *Natur prägt Kultur. Der Einfluß von Landschaft und Klima auf den Menschen: Zur Geschichte der Geophilosophie*. München: Telesma-Verlag.

Falter, R. (2009). Genius Monacensiae. Eine ortsbiografische Annäherung an den Genius Loci Münchens. In Mallien, L. and Heimrath, J. (eds), *Genius Loci. Der Geist von Orten und Landschaften in Geomantie und Architektur*. Klein Jasedow: Drachenverlag, pp. 84–105.

Ferrara, A. (2014). *The Democratic Horizon: Hyperpluralism and the Renewal of Political Liberalism*. New York: Cambridge University Press.

Fichte, J. G. (1992). *Foundations of Transcendental Philosophy (Wissenschaftslehre) Nova Methodo*. Ithaca, NY and London: Cornell University Press.

Fischer-Lichte, E. (2005). Zuschauen als Ansteckung. Katharsis oder Ansteckung. In Schaub, M., Suthor, N. and Fischer-Lichte, E. (eds), *Ansteckung. Zur Körperlichkeit eines ästhetischen Prinzips*. München: Fink, pp. 35–50.

Fischer-Lichte, E. (2008). *The Transformative Power of Performance: A New Aesthetics*. Oxon and New York: Routledge.

Ford, R. (2006). *The Lay of the Land*. Toronto: Knopf.

France, R. H. (1920). *Die Lebensgesetze einer Stadt*. München: Bruckmann.

Frese, J. (1995). 'Gefühls-Partituren'. In Großheim, M. (ed.), *Leib und Gefühl*. Berlin: Akademie Verlag, pp. 45–70.

Fridlund, A. J. (1994). *Human Facial Expression: An Evolutionary View*. San Diego, CA: Academic Press.

Friedrich, J. (2010). Das Erleben von Ausdruck – Einfühlung oder Zeichen? Zu Bühlers Ausdrucktheorie. In Grossheim, M. and Volke, S. (eds), *Gefühl, Geste, Gesicht. Zur Phänomenologie des Ausdrucks*. Freiburg-München: Alber, pp. 88–113.

Fromm, L. (2008). Wohnen in der Stadt. In Hasse, J. (ed.), *Die Stadt als Wohnraum*. Freiburg-München: Alber, pp. 63–93.

Früchtl, J. and Zimmermann, J. (2001). Ästhetik der Inszenierung. Dimensionen eines gesellschaftlichen, individuellen und kulturellen Phänomens. In Id. (ed.), *Ästhetik der Inszenierung*. Frankfurt a. M.: Suhrkamp, pp. 9–47.

Fuchs, T. (2000). *Leib, Raum, Person. Entwurf einer phänomenologischen Anthropologie*. Stuttgart: Klett-Cotta.

Führ, E. (1998). 'Genius loci'. Phänomen oder Phantom? *Wolkenkuckucksheim* 3, 2. www.cloud-cuckoo.net/openarchive/wolke/deu/Themen/982/Fuehr/fuehr_t.html

Funke, D. (2006). *Die dritte Haut. Psychoanalyse des Wohnens*. Gießen: Psychosozial Verlag.

Gallese, V. (2005). Embodied Simulation: From Neurons to Phenomenal Experience. *Phenomenology and Cognitive Science* 4, pp. 23–48.
Gamm, G. (1994). *Flucht aus der Kategorie*. Frankfurt a. M.: Suhrkamp.
Garnham, H. L. (1985). *Maintaining the Spirit of Place: A Process for the Preservation of Town Character*. Mesa, AZ: PDA Publishers.
Geiger, M. (1911). Zum Problem der Stimmungseinfühlung. *Zeitschrift für Ästhetik und allgemeine Kunstwissenschaft* 6, pp. 1–42.
Gibson, J. (1986). *The Ecological Approach to Visual Perception*. London: Routledge.
Giedion, S. (2002). *Space, Time and Architecture: The Growth of a New Tradition*. Cambridge, MA: Harvard University Press.
Giordano, G. (1997). *La casa vissuta. Percorsi e dinamiche dell'abitare*. Milano: Giuffrè Editore.
Goethe, J. W. (1981). Zur Geologie. In id., *Werke*, Hamburger edition. München: Beck, XIII, pp. 251–303.
Goethe, J. W. (1998). *Maxims and Reflections*. London: Penguin Books.
Goldie, P. (2000). *The Emotions: A Philosophical Exploration*. Oxford: Oxford University Press.
Graf, B. (2006). "Atmosphäre" in der Stadt – Denkansätze für einen psychologisch reflektierten Städtebau. *Wolkenkuckucksheim* X. www.cloud-cuckoo.net/openarchive/wolke/deu/Themen/051/Graf/graf.htm.
Grant, S. (2013). Performing an Aesthetics of Atmospheres. *Aesthetics* 23, 1, pp. 12–32.
Grassi, E. (1979). *Die Macht der Phantasie: zur Geschichte abendländischen Denkens*. Königstein: Athenäum.
Griffero, T. (1984). L'ermeneutica ricognitiva di E. D. Hirsch. *Rivista di estetica* XXV (16), pp. 77–91.
Griffero, T. (1988). *Interpretare. La teoria di Emilio Betti e il suo contesto*. Torino: Rosenberg & Sellier.
Griffero, T. (2006). *Il corpo spirituale. Ontologie 'sottili' da Paolo di Tarso a Friedrich Christoph Oetinger*. Milano: Mimesis.
Griffero, T. (2009). Vincoli situazionali. In Di Monte, M. and Rotili, M. (eds), *Vincoli/Constraints* (Sensibilia 2–2008). Milano: Mimesis, pp. 199–227.
Griffero, T. (2013). The Atmospheric 'Skin' of the City. *Ambiances. International Journal of Sensory Environment, Architecture and Urban Space*, pp. 1–19. http://ambiances.revues.org/399.
Griffero, T. (2014a). *Atmospheres: Aesthetics of Emotional Spaces*. London and New York: Routledge.
Griffero, T. (2014b). Atmospheres and Lived Space. *Studia Phaenomenologica* 14, pp. 29–51.
Griffero, T. (2014c). Who's Afraid of Atmospheres (And of Their Authority)? *Lebenswelt* IV (1), pp. 193–213.
Griffero, T. (2014d). Architectural Affordances: The Atmospheric Authority of Spaces. In Tidwell, P. (ed.), *Architecture and Atmosphere*. Espoo: Tapio Wirkkala-Rut Bryk Foundation, pp. 15–47.
Griffero, T. (2015). Fisiognomica emozionale. Affordances, estasi, atmosfere. *Lebenswelt* 6, pp. 53–79.
Griffero, T. (2016a). *Il pensiero dei sensi. Atmosfere ed estetica patica*. Milano: Guerini e Associati.
Griffero, T. (2016b). Urbilder o atmosfere? Ludwig Klages e la Nuova Fenomenologia. *Annuario Filosofico* 32, pp. 326–348.

Griffero, T. (2017a). *Quasi-Things: The Paradigm of Atmospheres.* New York: Suny Press.

Griffero, T. (2017b). Felt-Bodily Resonances: Towards a Pathic Aesthetics. *Yearbook for Eastern and Western Philosophy* 2, pp. 149–164.

Griffero, T. (2017c). Felt-bodily Communication: A Neophenomenological Approach to Embodied Affects. *Studi di estetica* XLV, IV, 2, pp. 71–86.

Griffero, T. (2018). Something more. Atmospheres and Pathic Aesthetics. In Griffero, T. and Moretti, G. (ed.), *Atmosphere/Atmospheres: Testing a New Paradigm*. Milan: Mimesis International, pp. 75–89.

Griffero, T. (2019). How do You Find Yourself in Your Environment? Hermann Schmitz's New Phenomenology? In Schmitz, H., *New Phenomenology: A Brief Introduction*. Milan: Mimesis International, pp. 9–39.

Griffiths, P. E. (1997). *What Emotions Really Are: The Problem of Psychological Categories.* Chicago, IL: University of Chicago Press.

Grossheim, M. (1994). *Ludwig Klages und die Phänomenologie.* Berlin: Akademie Verlag.

Grossheim, M., Kluck, S. and Nörenberg, H. (2014). Kollektive Lebensgefühle. Zur Phänomenologie von Gemeinschaften. *Rostocker Phænomenologische Manuskripte* 20.

Gumbrecht, H. U. (2003). *Production of Presence: What Meaning Cannot Convey.* Stanford, CA: Stanford University Press.

Gumbrecht, H. U. (2011). *Stimmungen lesen. Über eine verdeckte Wirklichkeit der Literatur.* München: Carl Hanser Verlag.

Gumbrecht, H. U. (2012a). *Präsenz*. Frankfurt a. M.: Suhrkamp.

Gumbrecht, H. U. (2012b). *Atmosphere, Mood, Stimmung: On a Hidden Potential of Literature.* Stanford, CA: Stanford University Press.

Gumbrecht, H. U. (2014). *Our Broad Present: Time and Contemporary Culture.* New York: Columbia University Press.

Günzel, S. (2014). Vom Raum zum Ort – und zurück. In Schlitte, A., Hünefeldt, T., Romic, D. and Loon, J.van (eds), *Philosophie des Ortes: Reflexionen zum Spatial Turn in den Sozial- und Kulturwissenschaften*. Bielefeld: [transcript], pp. 25–43.

Guzzoni, U. (1999). *Wohnen und Wandern.* Düsseldorf: ParErga.

Haapala, A. (2004). On the Aesthetics of the Everyday: Familiarity, Strangeness and the Meaning of Place. In Light, A. and Smith, J. M. (eds), *The Aesthetics of Everyday Life*. New York: Columbia University Press, pp. 39–55.

Haensel, C. (1946). *Das Wesen der Gefühle.* Überlingen: Werner Wulff Verlag.

Hahn, A. (2012a). Erlebnislandschaft−Erlebnis Landschaft? In Hahn, A. (ed.), *Erlebnislandschaft−Erlebnis Landschaft? Atmosphären im architektonischen Entwurf.* Bielefeld: [transcript], pp. 11–37.

Hahn, A. (2012b). Erlebnis Landschaft und das Erzeugen von Atmosphären. In Hahn, A. (ed.), *Erlebnislandschaft−Erlebnis Landschaft? Atmosphären im architektonischen Entwurf.* Bielefeld: [transcript], pp. 41–95.

Hall, E. T. (1966). *The Hidden Dimension*. New York: Anchor Books Doubleday.

Harrison, J. E. (1903). *Prolegomena to the Study of Greek Religion*. Cambridge: The University Press.

Hasse, J. (2000). *Die Wunden der Stadt. Für eine neue Ästhetik unserer Städte.* Wien: Passagen.

Hasse, J. (2005). *Fundsachen der Sinne. Eine phänomenologische Revision alltäglichen Erlebens.* Freiburg-München: Alber.

Hasse, J. (2008). Schöner wohnen? Zur Bedeutung von Ästhetisierung im Stadtraum. In Hasse, J. (ed.), *Die Stadt als Wohnraum*. Freiburg-München: Alber, pp. 109–132.

Hasse, J. (2009). *Unbedachtes Wohnen. Lebensformen an verdeckten Rändern der Gesellschaft*. Bielefeld: [transcript].

Hasse, J. (2012). *Atmosphären der Stadt. Aufgespürte Räume*. Berlin: jovis.

Hasse, J. (2014). *Was Räume mit uns machen – und wir mit ihnen. Kritische Phänomenologie des Raumes*. Freiburg-München: Alber.

Hasse, J. (2017). *Die Aura des Einfachen. Mikrologien räumlichen Erlebens*. 1. Freiburg-München: Alber.

Hauskeller, M. (1995). *Atmosphären erleben. Philosophische Untersuchungen zur Sinneswahrnehmung*. Berlin: Akademie Verlag.

Hauskeller, M. (2002). I could go for something Koons. Neue Ästhetik und Kommunikative Kunst. In Mahayni, Z. (ed.), *Neue Ästhetik. Das Atmosphärische und die Kunst*. München: Fink, pp. 173–182.

Hegel, G. W. F. (1975). *Aesthetics: Lectures on Fine Art*. Oxford: Clarendon Press.

Hegel, G. W. F. (2001). *Philosophy of Right*. Kitchener, Ontario: Batoche Books.

Heibach, C. (2012). Manipulative Atmosphären. Zwischen unmittelbarem Erleben und medialer Konstruktion. In id. (ed.), *Atmosphären. Dimensionen eines diffusen Phänomens*. München: Fink, pp. 261–282.

Heidegger, M. (1971). The Origin of the Work of Art. In Id., *Poetry, Language Thought*. New York: Harper & Row, pp. 15–86.

Heidegger, M. (1995). *The Fundamental Concepts of Metaphysics. World, Finitude, Solitude*. Bloomington, IN and Indianapolis, IN: Indiana University Press.

Heidegger, M. (2001). *Poetry, Language, Thought*. New York: Harper.

Hellpach, W. (1939). *Mensch und Volk der Grossstadt*. Stuttgart: Enke.

Hellpach, W. (1977). *Geopsyche. Die Menschenseele unter dem Einfluß von Wetter und Klima, Boden und Landschaft*. Stuttgart: Enke.

Heß, R. (2013). Stimmung, Atmosphäre, Präsenz. Wirkungsästhetische Begriffe zur Analyse der Architektur von Peter Zumthor. *Jahrbuch immersiver Medien*, pp. 55–66.

Heubel, F. (2003). Hermann Schmitz' Adolf Hitler in der Geschichte oder Zur Kritik der Neuen Phänomenologie. In Ogawa, T. (ed.), *Studies on New Phenomenology and Theories of Collective Consciousness*. Kyoto: Graduate School of Global Environmental Studies, pp. 41–51.

Hisayama, Y. (2014). *Erfahrungen des ki – Leibessphäre, Atmosphäre, Pansphäre*. Freiburg-München: Alber.

Hiss, T. (1990). *The Experience of Place*. New York: Random House.

Hoffmann, G. (1998). Das menschliche Sein als mediales. In Hauskeller, M., Rehmann-Sutter, C. and Schiemann, G. (eds), *Naturerkenntnis und Natursein. Für Gernot Böhme*. Frankfurt a. M.: Suhrkamp, pp. 338–354.

Home-Cook, G. (2015). *Theatre and Aural Attention: Stretching Ourselves*. Houndmills, Basingstoke, UK: Palgrave Macmillan.

Husserl, H. (2001). *Logical Investigations*. 2 vol. London: Routledge.

Husserl, E. (2002). Foundational Investigations of the Phenomenological Origin of the Spatiality of Nature: The Originary Ark, the Earth, Does Not Move. In Merleau-Ponty, M., *Husserl at the Limits of Phenomenology: Including Texts by Edmund Husserl*. Evanston, IL: Northwestern University Press, pp. 117–131.

Ingold, T. (2015). *The Life of Lines*. London and New York: Routledge.

Irrgang, B. (2016). Menschlicher Leib, personales Wissen, Praxis. Zur Kritik der Leibkonzeption der Neuen Phänomenologie. In Irrgang, B. and Rentsch, T. (eds),

"*Leib*" *in der neueren deutschen Philosophie*. Würzburg: Königshausen & Neumann, pp. 75–103.

Jäkel, A. (2013). *Gestik des Raumes. Zur leiblichen Kommunikation zwischen Benutzer und Raum in der Architektur*. Tübingen-Berlin: Wasmuth.

Jaschko, S. (2011). Performative Architektur–Mediale Erweiterungen und Dekonstruktionen von Räumen. In Lehnert, G. (ed.), *Raum und Gefühl. Der Spatial Turn und die neue Emotionsforschung*. Bielefeld: [transcript], pp. 109–117.

Kamleithner, C. (2009). Atmosphäre und Gebrauch. Zu zwei Grundbegriffen der Architekturästhetik. *Wolkenkuckucksheim* XIV. www.cloud-cuckoo.net/journal1996-2013/inhalt/de/heft/ausgaben/109/Kamleithner/kamleithner.php.

Kant, I. (2007). *Critique of Judgement*. Oxford and New York: Oxford University Press.

Kenny, A. (1963). *Action, Emotion and Will*. London: Routledge.

Kersting, W. (1992). Selbstbewußtsein, Zeitbewußtsein und zeitliche Wahrnehmung. Augustinus, Brentano, Husserl über das Hören von Melodien. In Forum f. Philosophie Bad Homburg (ed.), *Zeiterfahrung und Personalität*. Frankfurt a. M.: Suhrkamp, pp. 57–88.

Klages, L. (1929–32). *Der Geist als Widersacher der Seele*. 3 vol. Leipzig: Barth.

Klages, L. (1940). Einleitung. In Schüler, A., *Fragmente und Vorträge aus dem Nachlass*. Leipzig: Barth, pp. 1–119.

Klages, L. (1944). *Rhytmen und Runen*. Leipzig: Barth.

Klages, L. (1952). *Ursprünge der Seelenforschung*. Stutgart: Reclam.

Klages, L. (1964). *Sämtliche Werke*, 6, *Ausdruckskunde*. Bonn: Bouvier.

Klages, L. (1974). *Sämtliche Werke*, 3, *Philosophische Schriften*. Bonn: Bouvier.

Koffka, K. (1999). *Principles of Gestalt Psychology*. London: Routledge.

Kozljanič, R. J. (2004). *Der Geist eines Ortes. Kulturgeschichte und Phänomenologie des Genius Loci*. 2 vol., München: Albunea Verlag.

Kozljanič, R. J. (2010–11). Landschaft als physiognomisch-atmosphärisches und geistig-kulturelles Phänomen. In Hasse, J. and Kozlianič, R. J. (eds), *Gelebter, erfahrener und erinnerter Raum* (Jahrbuch für Lebensphilosophie, V). München: Albunea Verlag, pp. 158–176.

Kretzer, A. (2013). Architekturlehre und Atmosphäre. *Jahrbuch immersiver Medien*, pp. 117–123.

Kuhlmann, D. (1998). Der Geist des (W)ortes. *Wolkenkuckucksheim* 3, 2. www.cloud-cuckoo.net/openarchive/wolke/deu/Themen/982/Kuhlmann/kuhlmann_t.html.

Kümmerlen, R. (1929). *Zur Ästhetik bühnenräumlicher Prinzipien*. Diss., Ludwigsburg: Schmoll & Häussermann.

Kunz, H. (1966). Über vitale und intentionale Bedeutungsgehalte. In Baeyer, W.von and Griffith, R. M. (eds), *Conditio humana*. Berlin, Heidelberg and New York: Springer, pp. 162–200.

Landgrebe, L. (1974). Reflexionen zu Husserls Konstitutionslehre. *Tijdschrift voor filosofie* 36, pp. 466–482.

Landkammer, J. (2000). Homer bis Hitler. Die "Neue Phänomenologie" und die Versuchung der Geschichtsphilosophie. www.sicetnon.org/content/editormedien/schmitz.pdf.

Landweer, H. (2011). Affektive Intentionalität oder zentrierte Affektivität. In Slaby, J., Stephan, A., Walter, H. and Walter, S. (eds), *Affektive Intentionalität. Beiträge zur welterschließenden Funktion der menschlischen Gefühle*. Paderborn: mentis, pp. 230–251.

Latour, B. and Gagliardi, P. (eds). (2006). *Les atmosphères de la politique. Dialogue pour un monde commun*. Paris: Le Seuil.

Lauterbach, M. J. (2014). *"Gefühle mit der Autorität unbedingten Ernstes". Eine Studie zur religiösen Erfahrung in Auseinandersetzung mit Jürgen Habermas und Hermann Schmitz*. Freiburg-München: Alber.

Leatherbarrow, D. (2015). Atmospheric Conditions. In Steiner, H. and Sternberg, M. (eds), *Phenomenologies of the City: Studies in the History and the Philosophy of Architecture*. Burlington, VT: Ashgate, pp. 85–100.

Le Bon, G. (2002). *The Crowd: A Study of the Popular Mind*. Mineola, NY: Dover.

LeDoux, J. (1996). *The Emotional Brain (The Mysterious Underpinnings of Emotional Life)*. New York: Simon & Schuster.

Leeuw, G.van der, Smart, N. and Turner, J. E. (1955). *Religion in Essence and Manifestation*. Princeton, NJ: Princeton University Press.

Lehmann, H.-T. (1999). Die Gegenwart des Theaters. In Fischer-Lichte, E., Kolesch, D. and Weiler, C. (eds), *Transformationen, Theater der neunziger Jahre*. Berlin: Theater der Zeit, pp. 13–25.

Lehmann, H.-T. (2001). *Postdramatic Theatre*. London and New York: Routledge.

Lévinas, E. (1990). *Difficult Freedom: Essays on Judaism*. London: Athlone Press.

Lewin, K. (1926). Vorsatz, Wille und Bedürfnis. *Psychologische Forschungen* 7, pp. 294–385.

Lynch, K. (1960). *The Image of The City*. Cambridge, MA: MIT Press.

Mahayni, Z. (2003). *Feuer, Wasser, Erde, Luft. Eine Phänomenologie der Natur am Beispiel der vier Elemente*. Rostock: Koch.

Maldiney, H. (1966). Die Entdeckung der asthetischen Dimension in der Phanomenologie von Erwin Straus. In *Conditio Humana. Erwin W. Straus on his 75th birthday*. Berlin, Heidelberg and New York: Springer, pp. 210–232.

Mallgrave, H. F. (2013). *Architecture and Embodiment: The Implications of the New Sciences and Humanities for Design*. Oxon and New York: Routledge.

Malpas, J. (2004). *Place and Experience: A Philosophical Topography*. Cambridge: Cambridge University Press.

Marconi, D. (1988). Alcuni usi teorici della storia della filosofia. In Vattimo, G. (ed.), *Filosofia '87*. Roma-Bari: Laterza, pp. 21–40.

Marquard, O. (1989). *Farewell to Matters of Principle Philosophical Studies*. New York and Oxford: Oxford University Press.

Marquard, O. (1991). *In Defense of the Accidental. Philosophical Studies*. New York and Oxford: Oxford University Press.

Meerzon, Y. (2005). Body and Space: Michael Chekhov's Notion of Atmosphere as the Means of Creating Space in Theatre. *Semiotica* 155, 1/4, pp. 259–279.

Meisenheimer, W. (2006). *Das Denken des Leibes und der architektonische Raum*. Köln: König.

Meisenheimer, W. (2008). Die Konstruktion von Innenraumgefühlen durch Architektur. In Hasse, J. (ed.), *Die Stadt als Wohnraum*. Freiburg-München: Alber, pp. 40–43.

Merleau-Ponty, M. (2002). *Phenomenology of Perception*. London and New York: Routledge

Metzger, W. (1941). *Psychologie. Die Entwicklung ihrer Grundannahmen seit der Einführung des Experiments*. Dresden and Leipzig: Steinkopf.

Meuter, N. (2010). Die Universalität des Ausdrucks. Zur empirischen Grundlage eines anthropologischen Phänomens. In Grossheim, M. and Volke, S. (eds), *Gefühl, Geste, Gesicht. Zur Phänomenologie des Ausdrucks*. Freiburg-München: Alber, pp. 30–55.

Milev, Y. (2012). Design governance und breaking news: das Mediendesign der permanenten Katastrophe. In Heibach, C. (ed.), *Atmosphären. Dimensionen eines diffusen Phänomens.* München: Fink, pp. 285–303.

Minkowski, E. (1954). Espace, intimité, habitat. *Situation* I. Utrecht-Antwerpen: Spectrum, pp. 172–186.

Minkowski, E. (1970). *Lived Time: Phenomenological and Psychopathological Studies.* Evanston, IL: Northwestern University Press.

Mitscherlich, A. (1965). *Die Unwirtlichkeit unserer Städte: Anstiftung zum Unfrieden.* Frankfurt a. M.: Suhrkamp.

Müller, H.-J. (2009). Genius Loci und Genialogie. Der Genius Loci und sein anthropologischer Bezug für eine neue Praxis der Gestaltung von Identität. In Mallien, L. and Heimrath, J. (eds), *Genius Loci. Der Geist von Orten und Landschaften in Geomantie und Architektur.* Klein Jasedow: Drachenverlag, pp. 114–147.

Münkler, H. (2001). Die Theatralisierung der Politik. In Früchtl, J. and Zimmermann, J. (eds), *Ästhetik der Inszenierung.* Frankfurt a. M.: Suhrkamp, pp. 144–163.

Münsterberg, M. (ed.) (1916). *A Harvest of German Verse.* New York: Appleton & Co.

Musil, R. (1995). *The Man Without Qualities.* 2 vol. New York: Knopf.

Neddens, M. C. (1986). *Ökologisch orientierte Stadt- und Raumentwicklung: Genius loci, Leitbilder, Systemansatz, Planung. Eine integrierte Gesamtdarstellung.* Wiesbaden-Berlin: Bauverlag.

Nietzsche, F. (1999). *The Birth of Tragedy and Other Writings.* Cambridge and New York: Cambridge University Press.

Nietzsche, F. (2011). *Dawn. Thoughts on the Presumptions of Morality.* Stanford, CA: Stanford University Press.

Noe, A. (2009). *Out of Our Heads: Why You Are Not Your Brain, and Other Lessons from the Biology of Consciousness.* New York: Hill and Wang.

Norberg-Schulz, C. (1980). *Genius Loci: Towards a Phenomenology of Architecture.* New York: Rizzoli.

Norman, D. (2002). *The Design of Everyday Things.* New York: Doubleday Basic Books.

Norman, D. (2005). *Emotional Design: Why We Love (Or Hate) Everyday Things.* New York: Doubleday Basic Books.

Nowotny, H. (1995). *Eigenzeit. Entstehung und Strukturierung des Zeitgefühls.* Frankfurt a. M.: Suhrkamp.

Nussbaum, M. (2001). *Upheavals of Thought: The Intelligence of Emotions.* Cambridge: Cambridge University Press.

Ogawa, T. (2001). *Grund und Grenze des Bewusstseins. Interkulturelle Phänomenologie aus japanischer Sicht.* Würzburg: Königshausen & Neumann.

Packard, V. (2007). *The Hidden Persuaders.* New York: IG Publishing.

Pallasmaa, Y. (2005). *The Eyes of the Skin: Architecture and the Senses.* Chichester, UK: Wiley & Sons.

Pallasmaa, Y. (2011). Selfhood and the World: Lived Space, Vision and Hapticity. In Diaconu, M., Heuberger, E. and Mateus-Berr, R. (eds), *Senses and the City: An Interdisciplinary Approach to Urban Sensescapes.* Münster: Lit Verlag, pp. 49–62.

Pallasmaa, Y. (2014). Space, Place and Atmosphere: Emotion and Peripherical Perception in Architectural Experience. *Lebenswelt* 4, pp. 230–245.

Pennick, N. (2009). Anima Loci – Die Seele des Orts. In Mallien, L. and Heimrath, J. (eds), *Genius Loci. Der Geist von Orten und Landschaften in Geomantie und Architektur.* Klein Jasedow: Drachenverlag, pp. 61–83.

Pérez-Gómez, A. (2016). *Attunement: Architectural Meaning after the Crisis of Modern Science*. Cambridge and London: MIT Press.

Philippson, P. (1939). *Griechische Gottheiten in ihren Landschaften*. Oslo: A. W. Brugge.

Pinzer, D. (2013). Erlebniswelten und Technikphilosophie. In Hahn, A. (ed.), *Erlebnislandschaft–Erlebnis Landschaft? Atmosphären im architektonischen Entwurf*. Bielefeld: [transcript], pp. 97–120.

Poe, E. A. (1850–56). The Fall of the House of Usher. In *The Works of the Late Edgar Allan Poe*. The Griswold Edition, 4 vol. New York: Redfield. I, pp. 291–309.

Pratkanis, A. R. and Aronson, E. (2001). *Age of Propaganda: The Everyday Use and Abuse of Persuasion*. New York: Freeman.

Preusser, H.-P. (2015). *Pathische Ästhetik. Ludwig Klages und die Urgeschichte der Postmoderne*. Heidelberg: Winter.

Proust, M. (1992). *In Search of Lost Time*. I. *Swann's Way*. New York: Random House.

Raff, T. (2008). *Die Sprache der Materialien. Anleitung zu einer Ikonologie der Werkstoffe*. München: Deutscher Kunstverlag.

Rappe, G. (1995). *Archaische Leiberfahrung. Der Leib in der frühgriechischen Philosophie und in außereuropäischen Kulturen*. Berlin: Akademie Verlag.

Ratcliffe, M. (2008). *Feelings of Being: Phenomenology, Psychiatry and the Sense of Reality*. Oxford and New York: Oxford University Press.

Rauh, A. (2012). *Die besondere Atmosphäre. Ästhetische Feldforschungen*. Bielefeld: [transcript].

Rauh, A. (2018). *Concerning Astonishing Atmospheres: Aisthesis, Aura and Atmospheric Portfolio*. Milan: Mimesis International.

Reichhardt, U. (2009). *Die Rolle der Atmosphäre im Gestaltungs- und Planungsprozess. Eine empirische Untersuchung*. Hamburg: Verlag Dr. Kovač.

Relph, E. (1976). *Place and Placelessness*. London: Pion.

Relph, E. (2009). A Pragmatic Sense of Place. *Environmental and Architectural Phenomenology* 20, 3, pp. 24–31.

Ritter, J. (1974). *Subjektivität. Sechs Aufsätze*. Frankfurt a. M.: Suhrkamp.

Ritter, J. (2010). *Vorlesungen zur philosophischen Ästhetik (1947/48 und 1962)*. Göttingen: Wallstein.

Rodatz, C. (2010). *Der Schnitt durch den Raum. Atmosphärische Wahrnehmung in und außerhalb von Theaterräumen*. Bielefeld: [transcript].

Rorty, R. (1984). The Historiography of Philosophy: Four Genres. In Rorty, R., Schneewind, J. B. and Skinner, Q. (eds), *Philosophy in History: Essays on the Historiography of Philosophy*. Cambridge: Cambridge University Press, pp. 49–75.

Rykwert, J. (2000). *Seduction of Place: The History and Future of the City*. New York: Pantheon Books.

Salmela, M. (2006). True Emotions. *The Philosophical Quarterly* 56, 224, pp. 382–405.

Sartre, J.-P. (1969). *Being and Nothingness*. London: Routledge.

Sartre, J.-P. (2002). Intentionality: A Fundamental Idea of Husserl's Phenomenology. In Moran, D. and Mooney, T. (eds), *The Phenomenology Reader*. London: Routledge, pp. 382–384.

Schapp, W. (2004). *In Geschichten verstrickt. Zum Sein von Mensch und Ding*. Frankfurt a. M.: Klostermann.

von Scheve, C. and Salmela, M. (eds). (2014). *Collected Emotions: Perspectives from Psychology, Philosophy, and Sociology*. Oxford: Oxford University Press.

Schmidt-Lauber, B. (2003). *Gemütlichkeit. Eine kulturwissenschaftliche Annäherung*. Frankfurt and New York: campus.
Schmitt, C. (2003). *The Nomos of the Earth*. New York: Telos Press.
Schmitz, H. (1964). *System der Philosophie*, I. *Die Gegenwart*. Bonn: Bouvier.
Schmitz, H. (1965). *System der Philosophie*. II. 1. *Der Leib*. Bonn: Bouvier.
Schmitz, H. (1966). *System der Philosophie*. II. 2. *Der Leib im Spiegel der Kunst*. Bonn: Bouvier.
Schmitz, H. (1967). *System der Philosophie*. III. *Der Raum*. 1. *Der leibliche Raum*. Bonn: Bouvier.
Schmitz, H. (1968). *Subjektivität. Beiträge zur Phänomenologie und Logik*. Bonn: Bouvier.
Schmitz, H. (1969). *System der Philosophie*. III. 2. *Der Raum*. 2. *Der Gefühlsraum*. Bonn: Bouvier.
Schmitz, H. (1973). *System der Philosophie*. III. *Der Raum*, 3. *Der Rechtsraum*. Bonn: Bouvier.
Schmitz, H. (1977). *System der Philosophie*. III. 4. *Das Göttliche und der Raum*. Bonn: Bouvier.
Schmitz, H. (1978). *System der Philosophie*, Bd. III. 5, *Die Wahrnehmung*, Bonn: Bouvier.
Schmitz, H. (1980). *System der Philosophie*. IV. *Die Person*. Bonn: Bouvier.
Schmitz, H. (1989). *Leib und Gefühl. Materialien zu einer philosophischen Therapeutik*. Paderborn: Junfermann.
Schmitz, H. (1990). *Der unerschöpfliche Gegenstand. Grundzüge der Philosophie*. Bonn: Bouvier.
Schmitz, H. (1993). *Die Liebe*. Bonn: Bouvier.
Schmitz, H. (1994a). Wozu Neue Phänomenologie? In Großheim, M. (ed.), *Wege zu einer volleren Realität. Neue Phänomenologie in der Diskussion*. Berlin: Akademie Verlag, pp. 7–18.
Schmitz, H. (1994b). *Neue Grundlagen der Erkenntnistheorie*. Bonn: Bouvier.
Schmitz, H. (1996). *Husserl und Heidegger*. Bonn: Bouvier.
Schmitz, H. (1997). *Höhlengänge. Über die gegenwärtige Aufgabe der Philosophie*. Berlin: Akademie Verlag.
Schmitz, H. (1998). Situationen und Atmosphären. Zur Ästhetik und Ontologie bei Gernot Böhme. In Hauskeller, M., Rehmann-Sutter, C. and Schiemann, G. (eds), *Naturerkenntnis und Natursein. Für Gernot Böhme*. Frankfurt a. M.: Suhrkamp, pp. 176–190.
Schmitz, H. (1999a). *Adolf Hitler in der Geschichte*. Bonn: Bouvier.
Schmitz, H. (1999b). *Der Spielraum der Gegenwart*. Bonn: Bouvier.
Schmitz, H. (2002). (with Marx, G. and Moldzio, A.). *Begriffene Erfahrung. Beiträge zur antireduktionistischen Phänomenologie*. Rostock: Koch.
Schmitz, H. (2003). *Was ist Neue Phänomenologie?* Rostock: Koch.
Schmitz, H. (2005a). *Situationen und Konstellationen. Wider die Ideologie totaler Vernetzung*. Freiburg-München: Alber.
Schmitz, H. (2005b). *Im Dialog. Neun neugierige und kritische Fragen an die Neue Phänomenologie* (with W. Sohst). Berlin: Xenomoi.
Schmitz, H. (2005c). Was ist ein Phänomen? In Schmoll, D. and Kuhlmann, A. (eds), *Symptom und Phänomen. Phänomenologische Zugänge zum kranken Menschen*. Freiburg-München: Alber, pp. 16–28.
Schmitz, H. (2007). *Der Leib, der Raum und die Gefühle*. Bielefeld-Locarno: Sirius.

Schmitz, H. (2008a). *Logische Untersuchungen*. Freiburg-München: Alber.

Schmitz, H. (2008b). Die Legitimierbarkeit von Macht. In Wendel, H. J. and Kluck, S. (eds), *Zur Legitimierbarkeit von Macht*. Freiburg-München: Alber, pp. 5–19.

Schmitz, H. (2010). *Jenseits des Naturalismus*. Freiburg-München: Alber.

Schmitz, H. (2011). *Der Leib*. Berlin and Boston, MA: de Gruyter.

Schmitz, H. (2012). *Das Reich der Normen*. Freiburg-München: Alber.

Schmitz, H. (2013a). *Kritische Grundlegung der Mathematik. Eine phänomenologisch-logische Analyse*. Freiburg-München: Alber.

Schmitz, H. (2013b). Charisma. In Becker, H. (ed.), *Zugang zu Menschen. Angewandte Philosophie in zehn Berufsfeldern*. Freiburg-München: Alber, pp. 100–109.

Schmitz, H. (2014). *Atmosphären*. Freiburg-München: Alber.

Schmitz, H. (2015). *selbst sein. Über Identität, Subjektivität und Personalität*. Freiburg-München: Alber.

Schmitz, H. (2017). *Zur Epigenese der Person*. Freiburg-München: Alber.

Schmitz, H. (2019). *New Phenomenology: A Brief Introduction*. Milan: Mimesis International.

Schneider, K. (1959). *Clinical Psychopathology*. New York: Grune & Stratton.

Schouten, S. (2011). *Sinnliches Spüren. Wahrnehmung und Erzeugung von Atmosphären im Theater*. Berlin: Theater der Zeit.

Schouten, S. (2014). Dramaturgie der Atmosphären – Zur Wahnehmung inszenierter Stimmungsräume. In Weidinger, J. (ed.), *Atmosphären entwerfen*. Berlin: Universitätsverlag der TU Berlin, pp. 133–151.

Schrödl, J. (2009). Stimme und Emotion. Affektive Wirksamkeiten im postdramatischen Theater. *Forum Modernes Theater* 24, 2, pp. 169–182.

Schuler, A. (1997). *Cosmogonische Augen. Gesammelte Schriften*. Paderborn: Igel.

Schultheis, C. (1998). *Leiblichkeit – Kultur – Erziehung. Zur Theorie der elementaren Erziehung*. Weinheim: Deutscher Studien Verlag.

Schultheis, C. (2008). Macht und Erziehung. Überlegungen zur pathisch-leiblichen Dimension pädagogischen Handelns. In Wendel, H. J. and Kluck, S. (eds), *Zur Legitimierbarkeit von Macht*. Freiburg-München: Alber, pp. 99–115.

Schultheis, C. (2013). Nähe und Distanz in pädagogischen Beziehung. In Becker, H. (ed.), *Zugang zu Menschen. Angewandte Philosophie in zehn Berufsfeldern*. Freiburg-München: Alber, pp. 49–72.

Schürer, E. (2005). Provocation and Proclamation, Vision and Imagery: Expressionist Drama between German Idealism and Modernity. In Donahue, N. H. (ed.), *A Companion to the Literature of German Expressionism*. Rochester, NY: Camden House, pp. 231–254.

Seamon, D. (2000). A Way of Seeing People and Place: Phenomenology in Environment-Behavior Research. In Wapner, S., Demick, J., Yamamoto, C. T. and Minami, H. (eds), *Theoretical Perspectives in Environment-Behavior Research. Underlying Assumptions, Research Problems and Methodologies*. New York: Kluwer/Plenum, pp. 157–178.

Seamon, D. (2013). Place Attachment and Phenomenology: The Synergistic Dynamism of Place. In Manzo, L. C. and Devine-Wright, P. (eds), *Place Attachment. Advances in Theory, Methods and Applications*. London and New York: Routledge, pp. 11–22.

Seel, M. (2005). *Aesthetics of Appearing*. Stanford, CA: Stanford University Press.

Shirazi, M. R. (2008). 'Genius loci'. Phenomenology from Without. *Wolkenkuckucksheim* 12, 2. www.cloud-cuckoo.net/journal1996-2013/inhalt/en/issue/issues/207/Shirazi/shirazi.php.

Shusterman, R. (1999). Somaesthetics: A Disciplinary Proposal. *The Journal of Aesthetics and Art Criticism* 57, 3, pp. 299–313.

Shusterman, R. (2008). *Body Consciousness: A Philosophy of Mindfulness and Somaesthetics*. Cambridge: Cambridge University Press.

Shusterman, R. (2012). *Thinking Through the Body: Essays in Somaesthetics*. Cambridge: Cambridge University Press.

Simmel, G. (1993). The Alps. *Qualitative Sociology* 16, 2, pp. 179–184.

Sitte, C. (2006). *The Birth of Modern City Planning: With a Translation of the 1889 Austrian Edition of his City Planning According to Artistic Principles*. Mineola, NY: Dover Publications.

Slaby, J. (2007). Emotionaler Weltbezug. Ein Strukturschema im Anschluss an Heidegger. In Landweer, H. (ed.), *Gefühle – Struktur und Funktion*. Berlin: Akademie Verlag, pp. 93–112.

Slaby, J. (2014). Emotions and the extended mind. In Salmela, M. and Scheve, C.von (eds), *Collective Emotions*. Oxford: Oxford University Press, pp. 32–46.

Sloterdijk, P. (1987). *Critique of Cynical Reason*. Minneapolis, MN and London: Minnesota University Press.

Sloterdijk, P. (2001–2016). *Spheres*. 3 vol. Los Angeles: Semiotext(e).

Sloterdijk, P. (2009). *Philosophical Temperaments: From Plato to Foucault*. New York and Chichester: Columbia University Press.

Sloterdijk, P. (2012). *Terror from the Air*. Los Angeles, CA: Semiotext(e).

Soentgen, J. (1997). *Das Unscheinbare. Phänomenologische Beschreibungen von Stoffen, Dingen und fraktalen Gebilden*. Berlin: Akademie Verlag.

Soentgen, J. (1998). *Die verdeckte Wirklichkeit. Einführung in die Neue Phänomenologie von Hermann Schmitz*. Bonn: Bouvier.

Sparke, P. (2008). *The Modern Interior*. London: Reaktion Books.

Spitz, R. A. (1960). *Die Entstehung der ersten Objektbeziehungen*. Stuttgart: Klett.

Steele, F. (1981). *Sense of Place*. Boston: CBI Publishing.

Stepath, K. (2006). *Gegenwartskonzepte. Eine philosophisch-literaturwissenschaftliche Analyse temporaler Strukturen*. Würzburg: Köenigshausen & Neumann.

Stern, D. N. (1985). *The Interpersonal World of the Infant*. New York: Basic Books.

Straus, E. (1963). *The Primary World of Senses*. New York: Free Press of Glencoe.

Straus, E. (1980). *Phenomenological Psychology*. New York: Garland.

Stritzler-Levine, N. (1996). Three Visions of the Modern Home: Josef Frank, Le Corbusier and Alvar Aalto. In Frank, J. *Architect and Designer*. New Haven, CT and London: Yale University Press, pp. 253–259.

Ströker, E. (1977). *Philosophische Untersuchungen zum Raum*. Frankfurt a. M.: Klostermann.

Tanizaki, J. (1977). *In Praise of Shadows*. Stony Creek, CT: Leete's Island Books.

Tellenbach, H. (1968). *Geschmack und Atmosphäre*. Salzburg: Müller.

Thibaud, J.-P. (2003). Die sinnliche Umwelt von Städten. Zum Verständnis urbaner Atmosphären. In Hauskeller, M. (ed.), *Die Kunst der Wahrnehmung. Beiträge zu einer Philosophie der sinnlichen Erkenntnis*. Kusterdingen: Die Graue Edition, pp. 280–297.

Thibaud, J.-P. (2015). *En quête d'ambiances*. Gèneve: MetisPresses.

Thibaud, J.-P. and Thomas, R. (2004). L'ambiance comme expression de la vie urbaine. *Cosmopolitiques* 7, pp. 102–108.

Tidwell, P. (ed.). (2014). *Architecture and Atmosphere*. Espoo: Tapio Wirkkala-Rut Bryk Foundation.

Timm, H. (1993). Das Gesicht der Gegenwart. Vom Facialprinzip christlicher Religionsphänomenologie. In Großheim, M. and Waschkies, H.-J. (eds), *Rehabilitierung des Subjektiven. Festschrift für Hermann Schmitz*. Bonn: Bouvier, pp. 261–265.

Tuan, Y.-F. (1976). Geopiety: A Theme in Man's Attachment to Nature and to Place. In Lowenthal, H. and Bowden, M. J. (eds), *Geographies of the Mind: Essays in Historical Geography in Honour of J. K. Wright*. New York: Oxford University Press, pp. 11–39.

Tuan, Y.-F. (1977). *Space and Place: The Perspective of Experience*. London: Edward Arnold.

Valena, T. (2009). Zu einer Phänomenologie des Genius Loci. In Mallien, L. and Heimrath, J. (eds), *Genius Loci. Der Geist von Orten und Landschaften in Geomantie und Architektur*. Klein Jasedow: Drachenverlag, pp. 148–206.

Vazquez, D. (2010). *Manuale di psicogeografia*. Cuneo: Nerosubianco.

Vidler, A. (1992). *The Architectural Uncanny: Essays in the Modern Unhomely*. Cambridge, MA and London: MIT Press.

Vitta, M. (2008). *Dell'abitare. Corpi spazi oggetti immagini*. Torino: Einaudi.

Volke, S. (2007). *Sprachphysiognomik – Grundlagen einer leibphänomenologischen Beschreibung der Lautwahrnehmung*. Freiburg-München: Alber.

Volke, S. (2010). Wortgesichter – Zur physiognomischen Artikulation des Zeichenkörpers. In Grossheim, M. and Volke, S. (eds), *Gefühl, Geste, Gesicht. Zur Phänomenologie des Ausdrucks*. Freiburg-München: Alber, pp. 172–195.

Volkelt, J. (2013). *Phänomenologie und Metaphysik der Zeit*. Raleigh, NC: Lulu.

Wagner, M. (2001). *Das Material der Kunst. Eine andere Geschichte der Moderne*. München: Beck.

Waldenfels, B. (1996). Architektonik am Leitfaden des Leibes. *Wolkenkuckucksheim* I, 1. www.cloud-cuckoo.net/openarchive/wolke/deu/Themen/961/waldenfels/Waldenfels_t.html.

WaldenfelsB. (2000). *Das leibliche Selbst. Vorlesungen zur Phänomenologie des Leibes*. Frankfurt a. M.: Suhrkamp.

Warhol, A. (1975). *The Philosophy of Andy Warhol (From A to B and Back Again)*. New York and London: Harcourt Brace Jovanovich.

Warnke, M. (1979). Zur Situation der Couchecke. In Habermas, J. (ed.), *Stichworte zur «geistigen Situation der Zeit»*, II, *Politik und Kultur*. Frankfurt a. M.: Suhrkamp, pp. 673–687.

Warren, W. H.Jr. (1995). Self-motion: Visual Perception and Visual Control. In Epstein, W. and Rogers, S. (eds), *Perception of Space and Motion*. New York: Academic Press, pp. 263–325.

Watsuji, T. (1961). *Climate and Culture: A Philosophical Study*. Westport, CT: Greenwood Press.

Welsch, W. (1991). *La terra e l'opera d'arte. Heidegger e Il crepuscolo di Michelangelo*. Ferrara: Gallio.

Welton, M. (2012). *Feeling Theatre*. Houndmills, Basingstoke, UK: Palgrave Macmillan.

Werhahn, H. (ed.). (2011). *Neue Phänomenologie. Hermann Schmitz im Gespräch*. Freiburg-München: Alber.

White, G. (2012). On Immersive Theatre. *Theatre Research International* 37, 3, pp. 221–235.

Wiesing, L. (2014). *The Philosophy of Perception: Phenomenology and Image Theory*. London: Bloomsbury.

Wigley, M. (1998). The Architecture of Atmosphere. *Daidalos* 68, pp. 18–27.

Withagen, R., De Poel, H., Araujo, D. and Pepping, G.-J. (2012). Affordances Can Invite Behavior: Reconsidering the Relationship Between Affordances and Agency. *New Ideas in Psychology* 30, pp. 250–258.

Wolf, B. (2015). *Atmosphären des Aufwachsens*. Rostocker Phænomenologische Manuskripte 22.

Wölfflin, H. (1994). Prolegomena to a Psychology of Architecture. In Mallgrave, H. F. and Ikonomou, E. (eds), *Empathy, Form and Space: Problems in German Aesthetics, 1873–1893*. Santa Monica, CA: Getty Centre for the History of Art and the Humanities, pp. 149–190.

Wollheim, R. (2001). *On the Emotions*. New Haven, CT and London: Yale University Press.

Wydra, A. (2014). Intimate and Hostile Places: A Bachelardian Contribution to the Architecture of Lived Spaces. *Studia Phaenomenologica* 14, pp. 53–72.

Zaborowski, H. (2008). Zur Phänomenologie des Wohnens. In Hasse, J. (ed.), *Die Stadt als Wohnraum*. Freiburg-München: Alber, pp. 180–208.

Zanker, P. (1987). *The Power of Images in the Age of Augustus*. Ann Arbor, MI: University of Michigan Press.

Zoia, L. (2011). *Paranoia. La follia che fa la storia*. Turin: Bollati Boringhieri.

Zumthor, P. (1998). *Thinking Architecture*. Basel, Boston and Berlin: Birkhäuser.

Zumthor, P. (2006). *Atmosphären. Architektonische Umgebungen. Die Dinge um mich herum*. Basel, Boston and Berlin: Birkhäuser.

Index

For Product Safety Concerns and Information please contact our EU representative GPSR@taylorandfrancis.com
Taylor & Francis Verlag GmbH, Kaufingerstraße 24, 80331 München, Germany

www.ingramcontent.com/pod-product-compliance
Ingram Content Group UK Ltd.
Pitfield, Milton Keynes, MK11 3LW, UK
UKHW020007200926
475702UK00015B/2517
* 9 7 8 0 3 6 7 7 2 7 6 1 1 *